Monographs on Statistics and Applied Probability 108

Nonlinear Time Series
Semiparametric and
Nonparametric Methods

T0172738

MONOGRAPHS ON STATISTICS AND APPLIED PROBABILITY

General Editors

V. Isham, N. Keiding, T. Louis, S. Murphy, R. L. Smith, and H. Tong

Monographs on Statistics and Applied Probability 108

Nonlinear Time Series
Semiparametric and
Nonparametric Methods

Jiti Gao
The University of Western Australia
Perth, Australia

CRC Press
Taylor & Francis Group
Boca Raton London New York

CRC Press is an imprint of the
Taylor & Francis Group, an **informa** business
A CHAPMAN & HALL BOOK

CRC Press
Taylor & Francis Group
6000 Broken Sound Parkway NW, Suite 300
Boca Raton, FL 33487-2742

First issued in paperback 2019

© 2007 by Taylor & Francis Group, LLC
CRC Press is an imprint of Taylor & Francis Group, an Informa business

No claim to original U.S. Government works

ISBN-13: 978-1-58488-613-6 (hbk)
ISBN-13: 978-0-367-38935-2 (pbk)

Visit the Taylor & Francis Web site at
http://www.taylorandfrancis.com

and the CRC Press Web site at
http://www.crcpress.com

Contents

Preface

During the past two decades or so, there has been a lot of interest in both theoretical and empirical analysis of nonlinear time series data. Models and methods used have been based initially on parametric nonlinear or nonparametric time series models. Such parametric nonlinear models and related methods may be too restrictive in many cases. This leads to various nonparametric techniques being used to model nonlinear time series data. The main advantage of using nonparametric methods is that the data may be allowed to speak for themselves in the sense of determining the form of mathematical relationships between time series variables. In modelling nonlinear time series data one of the tasks is to study the structural relationship between the present observation and the history of the data set. The problem then is to fit a high dimensional surface to a nonlinear time series data set. While nonparametric techniques appear to be feasible and flexible, there is a serious problem: the so-called curse of dimensionality. For the independent and identically distributed case, this problem has been discussed and illustrated in the literature.

Since about twenty years ago, various semiparametric methods and models have been proposed and studied extensively in the economics and statistics literature. Several books and many papers have devoted their attention on semiparametric modelling of either independent or dependent time series data. The concentration has also been mainly on estimation and testing of both the parametric and nonparametric components in a semiparametric model. Interest also focuses on estimation and testing of conditional distributions using semiparametric methods. Important and useful applications include estimation and specification of conditional moments in continuous–time diffusion models. In addition, recent studies show that semiparametric methods and models may be applied to solve dimensionality reduction problems arising from using fully nonparametric models and methods. These include: (i) semiparametric single–index and projection pursuit modelling; (ii) semiparametric additive modelling; (iii) partially linear time series regression modelling; and (iv) semiparametric time series variable selection.

Although semiparametric methods in time series have recently been mentioned in several books, this monograph hopes to bring an up–to–date description of the recent development in semiparametric estimation, specification and selection of time series data as discussed in Chapters 1–4. In addition, semiparametric estimation and specification methods discussed in Chapters 2 and 3 are applied to a class of nonlinear continuous–time models with real data analysis in Chapter 5. Chapter 6 examines some newly proposed semiparametric estimation procedures for time series data with long–range dependence. While this monograph involves only climatological and financial data in Chapters 1 and 4–6, the newly proposed estimation and specifications methods are applicable to model sets of real data in many disciplines. This monograph can be used to serve as a textbook to senior undergraduate and postgraduate students as well as other researchers who are interested in the field of nonlinear time series using semiparametric methods.

This monograph concentrates on various semiparametric methods in model estimation, specification testing and selection of nonlinear time series data. The structure of this monograph is organized as follows: (a) Chapter 2 systematically studies estimation problems of various parameters and functions involved in semiparametric models. (b) Chapter 3 discusses parametric or semiparametric specification of various conditional moments. (c) As an alternative to model specification, Chapter 4 examines the proposed parametric, nonparametric and semiparametric model selection criteria to show how a time series data should be modelled using the best available model among all possible models. (d) Chapter 5 considers some of the latest results about semiparametric methods in model estimation and specification testing of continuous–time models. (e) Chapter 6 gives a short summary of recent semiparametric estimation methods for long–range dependent time series and then discusses some of the latest theoretical and empirical results using a so–called simultaneous semiparametric estimation method.

While the author of this monograph has tried his best to reflect the research work of many researchers in the field, some other closely related studies may be inevitably omitted in this monograph. The author therefore apologizes for any omissions.

I would like to thank anyone who has encouraged and supported me to finish the monograph. In particular, I would like to thank Vo Anh, Isabel Casas, Songxi Chen, Iréne Gijbels, Chris Heyde, Yongmiao Hong, Maxwell King, Qi Li, Zudi Lu, Peter Phillips, Peter Robinson, Dag Tjøstheim, Howell Tong and Qiying Wang for many helpful and stimulating discussions. Thanks also go to Manuel Arapis, Isabel Casas, Chaohua Dong, Kim Hawthorne and Jiying Yin for computing assistance as well as to Isabel Casas and Jiying Yin for editorial assistance. I would also like to acknowledge the generous support and inspiration of my colleagues in the School of Mathematics and Statistics at The University of Western Australia. Since the beginning of 2002, my research in the field has been supported financially by the Australian Research Council Discovery Grants Program.

My final thanks go to my wife, Mrs Qun Jiang, who unselfishly put my interest in the top priority while sacrificing hers in the process, for her constant support and understanding, and two lovely sons, Robert and Thomas, for their cooperation. Without such support and cooperation, it would not be possible for me to finish the writing of this monograph.

Jiti Gao
Perth, Australia
30 September 2006

CHAPTER 1

Introduction

1.1 Preliminaries

This monograph basically discusses semiparametric methods in model estimation, specification testing and selection of nonlinear time series data. We use the term *semiparametric* for models which are semiparametric partially linear models or other semiparametric regression models as discussed in Chapters 2–6, in particular Chapters 2 and 5. We also use the word *semiparametric* for methods which are semiparametric estimation and testing methods as discussed in Chapters 2–6, particularly in Chapters 3 and 6. Meanwhile, we also use the term *nonparametric* for models and methods which are either nonparametric models or nonparametric methods or both as considered in Chapters 2–5.

1.2 Examples and models

Let (Y, X) be a $d + 1$–dimensional vector of time series variables with Y being the response variable and X the vector of d–dimensional covariates. We assume that both X and Y are continuous random variables with $\pi(x)$ as the marginal density function of X, $f(y|x)$ being the conditional density function of Y given $X = x$ and $f(x, y)$ as the joint density function. Let $m(x) = E[Y|X = x]$ denote the conditional mean of Y given $X = x$. Let $\{(Y_t, X_t) : 1 \leq t \leq T\}$ be a sequence of observations drawn from the joint density function $f(x, y)$. We first consider a partially linear model of the form

$$Y_t = E[Y_t|X_t] + e_t = m(X_t) + e_t = U_t^\tau \beta + g(V_t) + e_t, \qquad (1.1)$$

where $X_t = (U_t^\tau, V_t^\tau)^\tau$, $m(X_t) = E[Y_t|X_t]$, and $e_t = Y_t - E[Y_t|X_t]$ is the error process and allowed to depend on X_t. In model (1.1), U_t and V_t are allowed to be two different vectors of time series variables. In practice, a crucial problem is how to identify U_t and V_t before applying model (1.1) to model sets of real data. For some cases, the identification problem can be solved easily by using empirical studies. For example, when modelling electricity sales, it is natural to assume the impact of temperature on electricity consumption to be nonlinear, as both high and low temperatures lead to increased consumption, whereas a linear relationship may be assumed for other regressors. See Engle *et al.* (1986). Similarly, when modelling the dependence of earnings on qualification and labour market experience variables, existing studies (see Härdle, Liang and Gao 2000) show that the impact of qualification on earnings to be linear, while the dependence of earnings on labour market experience appears to be nonlinear. For many other cases, however, the identification problem should be solved theoretically before using model (1.1) and will be discussed in detail in Chapter 4.

Existing studies show that although partially linear time series modelling may not be capable of reducing the nonparametric time series regression into a sum of one-dimensional nonparametric functions of individual lags, they can reduce the dimensionality significantly for some cases. Moreover, a feature of partially linear time series modelling is that it takes the true structure of the time series data into account and avoids neglecting some existing information on the linearity of the data.

We then consider a different partially linear model of the form

$$Y_t = X_t^\tau \beta + g(X_t) + e_t, \qquad (1.2)$$

where $X_t = (X_{t1}, \cdots, X_{td})^\tau$ is a vector of time series, $\beta = (\beta_1, \cdots, \beta_d)^\tau$ is a vector of unknown parameters, $g(\cdot)$ is an unknown function and can be viewed as a misspecification error, and $\{e_t\}$ is a sequence of either dependent errors or independent and identically distributed (i.i.d.) errors. In model (1.2), the error process $\{e_t\}$ is allowed to depend on $\{X_t\}$. Obviously, model (1.2) may not be viewed as a special form of model (1.1). The main motivation for systematically studying model (1.2) is that partially linear model (1.2) can play a significant role in modelling some nonlinear problems when the linear regression normally fails to appropriately model nonlinear phenomena. We therefore suggest using partially linear model (1.2) to model nonlinear phenomena, and then determine whether the nonlinearity is significant for a given data set (X_t, Y_t). In addition, some special cases of model (1.2) have already been considered in the econometrics and statistics literature. We show that several special forms of models (1.1) and (1.2) have some important applications.

We present some interesting examples and models, which are either special forms or extended forms of models (1.1) and (1.2).

Example 1.1 (Partially linear time series error models): Consider a partially linear model for trend detection in an annual mean temperature series of the form

$$Y_t = U_t^\tau \beta + g\left(\frac{t}{T}\right) + e_t, \qquad (1.3)$$

where $\{Y_t\}$ is the mean temperature series of interest, $U_t = (U_{t1}, \cdots, U_{tq})^\tau$ is a vector of q–explanatory variables, such as the southern oscillation index (SOI), t is time in years, β is a vector of unknown coefficients for the explanatory variables, $g(\cdot)$ is an unknown smooth function of time representing the trend, and $\{e_t\}$ represents a sequence of stationary time series errors with $E[e_t] = 0$ and $0 < \text{var}[e_t] = \sigma^2 < \infty$. Recently, Gao and Hawthorne (2006) have considered some estimation and testing problems for the trend function of the temperature series model (1.3).

Applying an existing method from Härdle, Liang and Gao (2000) to two global temperature series (http://www.cru.uea.ac.uk/cru/data/), Gao and Hawthorne (2006) have shown that a nonlinear trend looks feasible for each of the temperature series. Figure 1 of Gao and Hawthorne (2006) shows the annual mean series of the global temperature series from 1867–1993 and then from 1867–2001.

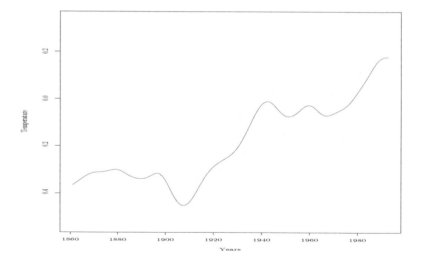

Figure 1.1 *The light line is the global temperature series for 1867–1993, while the solid curve is the estimated trend.*

Figure 1.1 shows that the trend estimate appears to be distinctly non-linear. Figure 1.2 displays the partially linear model fitting to the data set. The inclusion of the linear SOI component is warranted by the interannual fluctuations of the temperature series. Figures 1.1 and 1.2 also

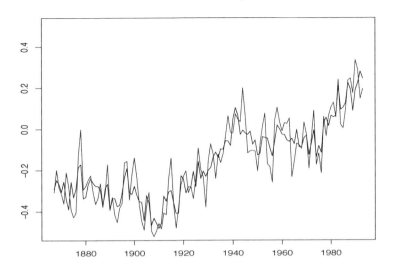

Figure 1.2 *The solid line is the global temperature series for 1867–1993, while the dashed line is the estimated series.*

show that the smooth trend component captures the nonlinear complexity inherent in the long term underlying trend. The mean function fitted to the data is displayed in Figure 1.3. The estimated series for the updated series is similar in stucture to that for the truncated series from 1867–1993. The hottest year on record, 1998, is represented reasonably. Similar to Figures 1.1 and 1.2, a kind of nonlinear complexity inherent in the long term trend is captured in Figure 1.3.

In addition, model (1.3) may be used to model long–range dependent (LRD) and nonstationary data. Existing studies show that there are both LRD and nonstationary properties inherited in some financial and environmental data (see Anh *et al.* 1999; Mikosch and Starica 2004) for example. Standard & Poor's 500 is a market–value weighted price of 500 stocks. The values in Figure 1.4 are from January 2, 1958 to July 29, 2005.

The key findings of such existing studies suggest that in order to avoid misrepresenting the mean function or the conditional mean function of a long–range dependent data, we should let the data 'speak' for themselves in terms of specifying the true form of the mean function or the conditional mean function. This is particularly important for data with

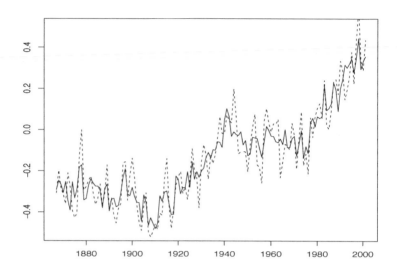

Figure 1.3 *The solid line is the global temperature series for 1867–2001, while the broken line is the estimated series.*

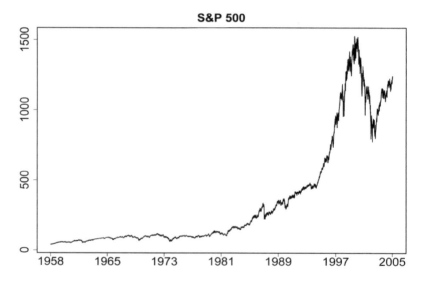

Figure 1.4 *S&P 500: January 2, 1958 to July 29, 2005.*

long–range dependence, because unnecessary nonlinearity or complexity in mean functions may cause erroneous LRD. Such issues may be addressed using a general model specification procedure to be discussed in Chapter 3 below.

Example 1.2 (Partially linear autoregressive models): Let $\{u_t\}$ be a sequence of time series variables, $Y_t = u_t$, $U_t = (u_{t-1}, \ldots, u_{t-q})^\tau$, and $V_t = (v_{t1}, \ldots, v_{tp})^\tau$ be a vector of time series variables. Now model (1.1) is a partially linear autoregressive model of the form

$$u_t = \sum_{i=1}^{q} \beta_i u_{t-i} + g(v_{t1}, \ldots, v_{tp}) + e_t. \tag{1.4}$$

When $\{v_t\}$ is a sequence of time series variables, $V_t = (v_{t-1}, \ldots, v_{t-p})^\tau$, $Y_t = v_t$, and $U_t = (u_{t1}, \ldots, u_{tq})^\tau$ be a vector of time series variables, model (1.1) is a partially nonlinear autoregressive model of the form

$$v_t = \sum_{i=1}^{q} \alpha_i u_{ti} + g(v_{t-1}, \ldots, v_{t-p}) + e_t. \tag{1.5}$$

In theory, various estimation and testing problems for models (1.4) and (1.5) have already been discussed in the literature. See for example, Robinson (1988), Tjøstheim (1994), Teräsvirta, Tjøstheim and Granger (1994), Gao and Liang (1995), Härdle, Lütkepohl and Chen (1997), Gao (1998), Härdle, Liang and Gao (2000), Gao and Yee (2000), and Gao, Tong and Wolff (2002a, 2002b), Gao and King (2005), and Li and Racine (2006).

In practice, models (1.4) and (1.5) have various applications. For example, Fisheries Western Australia (WA) manages commercial fishing in WA. Simple Catch and Effort statistics are often used in regulating the amount of fish that can be caught and the number of boats that are licensed to catch them. The establishment of the relationship between the Catch (in kilograms) and Effort (the number of days the fishing vessels spent at sea) is very important both commerically and ecologically. This example considers using a time series model to fit the relationship between catch and effort.

The historical monthly fishing data set from January 1976 to December 1999 available to us comes from the Fisheries WA Catch and Effort Statistics (CAES) database. Existing studies from the Fisheries suggest that the relationship between the catch and the effort does not look linear while the dependence of the current catch on the past catch appears to be linear. This suggests using a partially linear model of the form

$$C_t = \beta_1 C_{t-1} + \ldots + \beta_q C_{t-q} + g(E_t, E_{t-1}, \ldots, E_{t-p+1}) + e_t, \tag{1.6}$$

where $\{e_t\}$ is a sequence of random errors, C_t and E_t represent the catch and the effort at time t, respectively, and $g(\cdot)$ is a nonlinear function. In the detailed computation, we use the transformed data $Y_t = \log_{10}(C_t)$ and $X_t = \log_{10}(E_t)$ satisfying the following model

$$Y_{t+r} = \beta_1 Y_{t+r-1} + \ldots + \beta_q Y_{t+r-q} + g(X_{t+r}, \ldots, X_{t+r-p+1}) + e_t, \quad (1.7)$$

where $r = \max(p, q)$ and $\{e_t\}$ is a random error with zero mean and finite variance.

Gao and Tong (2004) proposed a semiparametric variable selection procedure for model (1.1) and then applied the proposed semiparametric selection method to produce the corresponding plots in Figure 1 of their paper.

Model (1.1) also covers the following important classes of partially linear time series models as given in Example 1.3 below.

Example 1.3 (Population biology model): Consider a partially linear time series model of the form

$$Y_t = \beta Y_{t-1} + g(Y_{t-\tau}) + e_t, \quad (1.8)$$

where $|\beta| < 1$ is an unknown parameter, $g(\cdot)$ is a smooth function such that $\{Y_t\}$ is strictly stationary, $\tau \geq 2$ is an integer, and $\{e_t\}$ is a sequence of strictly stationary errors. When $g(x) = \frac{bx}{1+x^k}$, we have a population biology model of the form

$$Y_t = \beta Y_{t-1} + \frac{bY_{t-\tau}}{1 + Y_{t-\tau}^k} + e_t, \quad (1.9)$$

where $0 < \beta < 1$, $b > 0$, $\tau > 1$ and $k \geq 1$ are parameters. The motivation for studying this model stems from the research of population biology model and the Mackey–Glass system. The idea of a threshold is very natural to the study of population biology because the production of eggs (young) per adult per season is generally a saturation–type function of the available food and food supply is generally limited. Here $\{Y_t\}$ denotes the number of adult flies in day t, a is the daily adult survival rate, d is the time delay between birth and maturation, and $\frac{bY_{t-\tau}}{1+Y_{t-\tau}^k}$ accounts for the recruitment of new adults due to births d years in the past, which is nonlinear because of decreased fecundity at higher population levels. Such a class of models have been discussed in Gao (1998) and Gao and Yee (2000).

Example 1.4 (Environmetric model): Consider a partially linear model of the form

$$Y_t = \sum_{i=1}^{q} \beta_i Y_{t-i} + g(V_t) + e_t, \quad (1.10)$$

where $\{Y_t\}$ denotes the air quality time series at t period, and $\{V_t\}$ represents a vector of many important factors such as wind speed and temperature. When choosing a suitable vector for $\{V_t\}$, we need to take all possible factors into consideration on the one hand but to avoid the computational difficulty caused by the spareness of the data and to provide more precise predictions on the other hand. Thus, for this case only wind speed, temperature and one or two other factors are often selected as the most significant factors. Such issues are to be addressed in Chapter 4 below.

When the dimension of $\{V_t\}$ is greater than three, we may suggest using a partially linear additive model of the form

$$Y_t = \sum_{i=1}^{q} \beta_i Y_{t-i} + \sum_{j=1}^{p} g_j(V_{tj}) + e_t, \qquad (1.11)$$

where each $g_j(\cdot)$ is an unknown function defined over $\boldsymbol{R}^1 = (-\infty, \infty)$. Model estimation, specification and selection for models in Examples 1.1–1.4 are to be discussed in Chapters 2–4 below.

Example 1.5 (Semiparametric single–index model): Consider a generalized partially linear time series model of the form

$$Y_t = X_t^{\tau} \theta + \psi(X_t^{\tau} \eta) + e_t, \qquad (1.12)$$

where (θ, η) are vectors of unknown parameters, $\psi(\cdot)$ is an unknown function over $\boldsymbol{R}^1$, and $\{e_t\}$ is a sequence of errors. The parameters and function are chosen such that model (1.12) is identifiable. While model (1.12) imposes certain additivity conditions on both the parametric and nonparametric components, it has been shown to be quite efficient for modelling high–dimensional time series data. Recent studies include Carroll *et al.* (1997), Gao and Liang (1997), Xia, Tong and Li (1999), Xia *et al.* (2004), and Gao and King (2005).

In recent years, some other semiparametric time series models have also been discussed as given below.

Example 1.6 (Semiparametric regression models): Consider a linear model with a nonparametric error model of the form

$$Y_t = X_t^{\tau} \beta + u_t \quad \text{with} \quad u_t = g(u_{t-1}) + \epsilon_t, \qquad (1.13)$$

where X_t and β are p–dimensional column vectors, $\{X_t\}$ is stationary with finite second moments, Y_t and u_t are scalars, $g(\cdot)$ is an unknown function and possibly nonlinear, and is such that $\{u_t\}$ is at least stationary with zero mean and finite variance i.i.d. innovations ϵ_t. Model (1.13) was proposed by Hidalgo (1992) and then estimated by a kernel-based procedure.

Truong and Stone (1994) considered a nonparametric regression model with a linear autoregressive error model of the form

$$Y_t = g(X_t) + u_t \quad \text{with} \quad u_t = \theta u_{t-1} + \epsilon_t, \qquad (1.14)$$

where $\{(X_t, Y_t)\}$ is a bivariate stationary time series, θ, satisfying $|\theta| < 1$, is an unknown parameter, $g(\cdot)$ is an unknown function, and $\{\epsilon_t\}$ is a sequence of independent errors with zero mean and finite variance $0 < \sigma^2 < \infty$. Truong and Stone (1994) proposed a semiparametric estimation procedure for model (1.14).

Example 1.7 (Partially linear autoregressive conditional heteroscedasticity (ARCH) models): For the case where $d = 1$, $\{Y_t\}$ is a time series, $X_t = Y_{t-1}$, and $\{e_t\}$ depends on Y_{t-1}, model (1.2) is a partially linear ARCH model of the form

$$Y_t = \beta Y_{t-1} + g(Y_{t-1}) + e_t, \qquad (1.15)$$

where $\{e_t\}$ is assumed to be stationary, both β and g are identifiable, and $\sigma^2(y) = E[e_t^2 | Y_{t-1} = y]$ is a smooth function of y. Hjellvik and Tjøstheim (1995), and Hjellvik, Yao and Tjøstheim (1998), Li (1999), and Gao and King (2005) all considered testing for linearity in model (1.15). Granger, Inoue and Morin (1997) have considered some estimation problems for the case of $\beta = 1$ in model (1.15).

Example 1.8 (Nonlinear and nonstationary time series models): This example considers two classes of nonlinear and nonstationary time series models. The first class of models is given as follows:

$$Y_t = m(X_t) + e_t \quad \text{with} \quad X_t = X_{t-1} + \epsilon_t, \qquad (1.16)$$

where $\{\epsilon_t\}$ is a sequence of stationary errors. The second class of models is defined by

$$Y_t = Y_{t-1} + g(Y_{t-1}) + e_t. \qquad (1.17)$$

Recently, Granger, Inoue and Morin (1997) considered the case where $g(\cdot)$ of (1.17) belongs to a class of parametric nonlinear functions and then discussed applications in economics and finance. In nonparametric kernel estimation of $m(\cdot)$ in (1.16) and $g(\cdot)$ of (1.17), existing studies include Karlsen and Tjøstheim (1998), Phillips and Park (1998), Karlsen and Tjøstheim (2001), and Karlsen, Myklebust and Tjøstheim (2006). The last paper provides a class of nonparametric versions of some of those parametric models proposed in Engle and Granger (1987). Model (1.16) corresponds to a class of parametric nonlinear models discussed in Park and Phillips (2001).

Compared with nonparametric kernel estimation, nonparametric specification testing problems for models (1.16) and (1.17) have just been considered in Gao *et al.* (2006). Specifically, the authors have proposed a novel unit root test procedure for stationarity in a nonlinear time series setting. Such a test procedure can initially avoid misspecification through the need to specify a linear conditional mean. In other words, the authors have considered estimating the form of the conditional mean and testing for stationarity simultaneously. Such a test procedure may also be viewed as a nonparametric counterpart of those tests proposed in Dickey and Fuller (1979), Phillips (1987) and many others in the parametric linear time series case.

Example 1.9 (Semiparametric diffusion models): This example involves using model (1.2) to approximate a continuous-time process of the form

$$dr_t = \mu(r_t)dt + \sigma(r_t)dB_t, \tag{1.18}$$

where $\mu(\cdot)$ and $\sigma(\cdot)$ are respectively the drift and volatility functions of the process, and B_t is standard Brownian motion. Since there are inconsistency issues for the case where both $\mu(\cdot)$ and $\sigma(\cdot)$ are nonparametric, we are mainly interested in the case where one of the functions is parametric. The first case is where $\mu(r, \theta)$ is a known parametric function indexed by a vector of unknown parameters, $\theta \in \Theta$ (a parameter space), and $\sigma(r)$ is an unknown but sufficiently smooth function.

The main motivation for considering such a class of semiparametric diffusion models is due to: (a) most empirical studies suggest using a simple form for the drift function, such as a polynomial function; (b) when the form of the drift function is unknown and sufficiently smooth, it may be well–approximated by a parametric form, such as by a suitable polynomial function; (c) the drift function may be treated as a constant function or even zero when interest is on studying the stochastic volatility of $\{r_t\}$; and (d) the precise form of the diffusion function is very crucial, but it is quite problematic to assume a known form for the diffusion function due to the fact that the instantaneous volatility is normally unobservable. The second case is where $\sigma(r, \vartheta)$ is a positive parametric function indexed by a vector of unknown parameters, $\vartheta \in \Theta$ (a parameter space), and $\mu(r)$ is an unknown but sufficiently smooth function. As pointed out in existing studies, such as Kristensen (2004), there is some evidence that the assumption of a parametric form for the diffusion function is also reasonable in such cases where the diffusion function is already pre–specified, the main interest is, for example, to specify whether the drift function should be linear or quadratic.

Model (1.18) has been applied to model various economic and financial data sets, including the two popular interest rate data sets given in Figures 1.5 and 1.6.

Recently, Arapis and Gao (2006) have proposed some new estimation

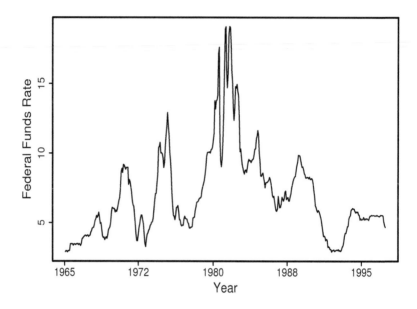

Figure 1.5 *Three-month T-Bill rate, January 1963 to December 1998.*

and testing procedures for model (1.18) using semiparametric methods. Such details, along with some other recent developments, are discussed in Chapter 5 below.

Example 1.10 (Continuous–time models with long–range dependence): Recent studies show that the standard Brownian motion involved in (1.18) needs to be replaced by a fractional Brownian motion when data exhibit long–range dependence. Comte and Renault (1996, 1998) proposed using a continuous–time model of the form

$$dZ(t) = -\alpha Z(t)dt + \sigma dB_\beta(t), \ \ Z(0) = 0, \ t \in (0, \infty), \tag{1.19}$$

where $B_\beta(t)$ is general fractional Brownian motion given by $B_\beta(t) = \int_0^t \frac{(t-s)^\beta}{\Gamma(1+\beta)} dB(s)$, and $\Gamma(x)$ is the usual Γ function. Gao (2004) then discussed some estimation problems for the parameters involved. More recently, Casas and Gao (2006) have systematically established both large and finite sample results for such estimation problems. Some of these results are discussed in Chapter 6 below.

More recently, Casas and Gao (2006) have proposed a so–called simul-

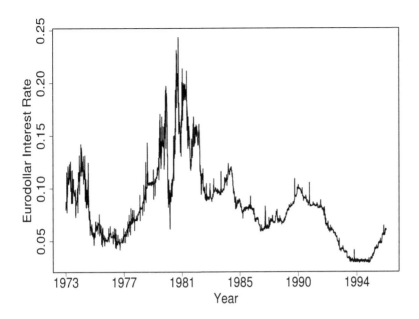

Figure 1.6 *Seven-Day Eurodollar Deposit rate, June 1, 1973 to February 25, 1995.*

taneous semiparametric estimation procedure for a class of stochastic volatility models of the form

$$dY(t) = V(t)dB_1(t) \quad \text{and} \quad dZ(t) = -\alpha Z(t)dt + \sigma dB_\beta(t), \quad (1.20)$$

where $V(t) = e^{Z(t)}$, $Y(t) = \ln(S(t))$ with $S(t)$ being the return process, $B_1(t)$ is a standard Brownian motion and independent of $B(t)$. The paper by Casas and Gao (2006) has established some asymptotic theory for the proposed estimation procedure. Both the proposed theory and the estimation procedure are illustrated using simulated and real data sets, including the S&P 500 data. To show why the S&P 500 data may show some kind of long–range dependence, Table 1.1 provides autocorrelation values for several versions of the compounded returns of the S&P 500 data.

Chapter 6 below discusses some details about both the estimation and implementation of model (1.20).

data	lag1	2	5	10	20	40	70	100
$T = 500$								
W_t	0.0734	-0.0458	0.0250	0.0559	-0.0320	-0.0255	0.0047	0.0215
$|W_t|^{1/2}$	-0.0004	0.1165	0.1307	0.0844	0.0605	-0.0128	0.0430	-0.0052
$|W_t|$	0.0325	0.1671	0.1575	0.1293	0.092	-0.0141	0.0061	-0.0004
$|W_t|^2$	0.0784	0.2433	0.1699	0.1573	0.1117	-0.0094	-0.0283	0.0225
$T = 2000$								
W_t	0.0494	-0.0057	-0.0090	0.0142	0.0012	-0.0209	0.0263	0.0177
$|W_t|^{1/2}$	-0.0214	-0.0072	0.0826	0.0222	0.0280	-0.0040	0.0359	0.0001
$|W_t|$	-0.0029	0.0187	0.0997	0.0258	0.0505	0.0036	0.0422	-0.0020
$|W_t|^2$	0.0401	0.0562	0.1153	0.0275	0.0668	0.0018	0.0376	-0.0045
$T = 10000$								
W_t	0.1580	-0.0224	0.0122	0.0125	0.0036	0.0079	0.0028	0.0071
$|W_t|^{1/2}$	0.1161	0.0813	0.1196	0.0867	0.0789	0.0601	0.0775	0.0550
$|W_t|$	0.1223	0.0986	0.1326	0.0989	0.0944	0.0702	0.0879	0.0622
$|W_t|^2$	0.1065	0.1044	0.1281	0.0937	0.0988	0.0698	0.0847	0.0559
$T = 16127$								
W_t	0.0971	-0.0362	0.0054	0.0180	0.0036	0.0222	-0.0061	0.0041
$|W_t|^{1/2}$	0.1783	0.1674	0.1879	0.1581	0.1567	0.1371	0.1252	0.1293
$|W_t|$	0.2044	0.2012	0.2215	0.1831	0.1835	0.1596	0.1439	0.1464
$|W_t|^2$	0.1864	0.2018	0.2220	0.1684	0.1709	0.1510	0.1303	0.1321

Table 1.1 Autocorrelation of W_t, $|W|^\rho$ for $\rho = \frac{1}{2}, 1, 2$ for the S&P 500 where $W_t = \ln\left(\frac{S_t}{S_{t-1}}\right)$ with $\{S_t\}$ be the S&P 500 daily values.

1.3 Bibliographical notes

Recent books on parametric linear and nonlinear time series include Tong (1990), Granger and Teräsvirta (1993), Tanaka (1996), Franses and Van Dijk (2000), Galka (2000), Chan (2002), Fan and Yao (2003), Kantz and Schreiber (2004), Tsay (2005), and Granger, Teräsvirta and Tjøstheim (2006).

In addition, nonparametric methods have been applied to model both independent and dependent time series data as discussed in Fan and Gijbels (1996), Hart (1997), Eubank (1999), Pagan and Ullah (1999), Fan and Yao (2003), Granger, Teräsvirta and Tjøstheim (2006), and Li and Racine (2006).

Applications of semiparametric methods and models to time series data have been discussed in Fan and Gijbels (1996), Pagan and Ullah (1999), Härdle, Liang and Gao (2000), Fan and Yao (2003), Ruppert, Wand and Carroll (2003), Granger, Teräsvirta and Tjøstheim (2006), and Li and Racine (2006).

CHAPTER 2

Estimation in Nonlinear Time Series

2.1 Introduction

This chapter considers semiparametric modelling of nonlinear time series data. We first propose an additive partially linear modelling method. A semiparametric single–index modelling procedure is then considered. Both new estimation methods and implementation procedures are discussed in some detail. The main ideas are to use either a partially linear form or a semiparametric single–index form to approximate the conditional mean function rather than directly assuming that the true conditional mean function is of either a partially linear form or a semiparametric single–index form.

2.1.1 Partially linear time series models

In time series regression, nonparametric methods have been very popular both for prediction and characterizing nonlinear dependence. Let $\{Y_t\}$ and $\{X_t\}$ be the one–dimensional and d–dimensional time series data, respectively. For a vector of time series data $\{Y_t, X_t\}$, the conditional mean function $E[Y_t|X_t = x]$ of Y_t on $X_t = x$ may be estimated nonparametrically by the Nadaraya–Watson (NW) estimator when the dimensionality d is less than three. When d is greater than three, the conditional mean can still be estimated using the NW estimator, and an asymptotic theory can be constructed. In practice, however, because of the so–called curse of dimensionality, this may not be recommended unless the number of data points is extremely large.

There are several ways of circumventing the curse of dimensionality in time series regression. Perhaps the two most commonly used are semiparametric additive models and single–index models. In time series regression, semiparametric additive fitting can be thought of as an approximation of conditional quantities such as $E[Y_t|Y_{t-1}, \ldots, Y_{t-d}]$, and sometimes (Sperlich, Tjøstheim and Yang 2002) interaction terms are included to improve this approximation. An advantage of using the semi-

15

parametric additive approach is that a priori information concerning possible linearity of some of the components can be included in the model. More specifically, we will look at approximating the conditional mean function $m(X_t) = m(U_t, V_t) = E[Y_t | U_t, V_t]$ by a semiparametric (partially linear) function of the form

$$m_1(U_t, V_t) = \mu + U_t^\tau \beta + g(V_t) \qquad (2.1)$$

such that $E[Y_t - m_1(U_t, V_t)]^2$ is minimized over a class of semiparametric functions of the form $m_1(U_t, V_t)$ subject to $E[g(V_t)] = 0$ for the identifiability of $m_1(U_t, V_t)$, where μ is an unknown parameter, $\beta = (\beta_1, \ldots, \beta_q)^\tau$ is a vector of unknown parameters, $g(\cdot)$ is an unknown function over $\mathbf{R}^p$, both $U_t = (U_{t1}, \ldots, U_{tq})^\tau$ and $V_t = (V_{t1}, \ldots, V_{tp})^\tau$ may be vectors of time series variables.

Motivation for using the form (2.1) for independent data analysis can be found in Härdle, Liang and Gao (2000). As for the independent data case, estimating $g(\cdot)$ in model (2.1) may suffer from the curse of dimensionality when $g(\cdot)$ is not necessarily additive and $p \geq 3$. Thus, this chapter proposes two different estimation methods. The first estimation method deals with the case where $m(x)$ is itself an additive partially linear form and each of the nonparametric components is approximated by a series of orthogonal functions. For the independent data case, the orthogonal series estimation method has been used as an alternative to some other nonparametric estimation methods, such as the kernel method. Recent monographs include Eubank (1999). As shown in Gao, Tong and Wolff (2002a), this method provides some natural parametric approximations to additive partially linear forms.

2.1.2 Semiparametric additive time series models

The main ideas of proposing the second method are taken from Gao, Lu and Tjøstheim (2006), who have established an estimation procedure for semiparametric spatial regression. The second method applies to the case where $m(x)$ is approximated by (2.1) and then proposes approximating $g(\cdot)$ by $g_a(\cdot)$, an additive marginal integration projector as detailed in the following section. When $g(\cdot)$ itself is additive, i.e., $g(x) = \sum_{i=1}^p g_i(x_i)$, the form of $m_1(U_t, V_t)$ can be written as

$$m_1(U_t, V_t) = \mu + U_t^\tau \beta + \sum_{i=1}^p g_i(V_{ti}) \qquad (2.2)$$

subject to $E[g_i(V_{ti})] = 0$ for all $1 \leq i \leq p$ for the identifiability of $m_1(U_t, V_t)$ in (2.2), where $g_i(\cdot)$ for $1 \leq i \leq p$ are all unknown one–dimensional functions over $\mathbf{R}^1$.

Our method of estimating $g(\cdot)$ or $g_a(\cdot)$ is based on an additive marginal integration projection on the set of additive functions, but where unlike the backfitting case, the projection is taken with the product measure of V_{tl} for $l = 1, \cdots, p$ (Nielsen and Linton 1998). This contrasts with the smoothed backfitting approach of Mammen, Linton and Nielsen (1999) to the nonparametric regression case. Marginal integration, although inferior to backfitting in asymptotic efficiency for purely additive models, seems well suited to the framework of partially linear estimation. In fact, in previous work (Fan, Härdle and Mammen 1998; Fan and Li 2003, for example) in the independent regression case marginal integration has been used, and we do not know of any work extending the backfitting theory to the partially linear case. Marginal integration techniques are also applicable to the case where interactions are allowed between the the V_{tl}–variables (cf. also the use of marginal integration for estimating interactions in ordinary regression problems).

2.1.3 Semiparametric single–index models

As an alternative to (2.2), we assume that $m(x) = E[Y_t|X_t = x] = m_2(X_t)$ is given by the semiparametric single–index form

$$m_2(X_t) = X_t^\tau \theta + \psi(X_t^\tau \eta). \tag{2.3}$$

When we partition $X_t = (U_t^\tau, V_t^\tau)^\tau$ and take $\theta = (\beta^\tau, 0, \cdots, 0)^\tau$ and $\eta = (0, \cdots, 0, \alpha^\tau)^\tau$, form (2.3) becomes the generalized partially linear form

$$m_2(X_t) = U_t^\tau \beta + \psi(V_t^\tau \alpha). \tag{2.4}$$

Various versions of (2.3) and (2.4) have been discussed in the econometrics and statistics literature. Recent studies include Härdle, Hall and Ichimura (1993), Carroll et al. (1997), Gao and Liang (1997), Xia, Tong and Li (1999), and Gao and King (2005).

In Sections 2.2 and 2.3 below, some detailed estimation procedures for $m_1(U_t, V_t)$ and $m_2(X_t)$ are proposed and discussed extensively. Section 2.2 first assumes that the true conditional mean function $m(x) = E[Y_t|X_t = x]$ is of the form (2.2) and develops an orthogonal series estimation method for the additive form. Section 2.3 then proposes an additive marginal integration projection method to estimate form (2.1) without necessarily assuming the additivity in (2.2).

2.2 Semiparametric series estimation

In this section, we employ the orthogonal series method to estimate each nonparametric function in (2.2). By approximating each $g_i(\cdot)$ by an orthogonal series $\sum_{j=1}^{n_i} f_{ij}(\cdot)\theta_{ij}$ with $\{f_{ij}(\cdot)\}$ being a sequence of orthogonal functions and $\{n_i\}$ being a sequence of positive integers, we have an approximate model of the form

$$Y_t = \mu + U_t^\tau \beta + \sum_{i=1}^{p} \sum_{j=1}^{n_i} f_{ij}(V_{ti})\theta_{ij} + e_t, \tag{2.5}$$

which covers some natural parametric time series models. For example, when $U_{tl} = U_{t-l}$ and $V_{ti} = Y_{t-i}$, model (2.5) becomes a parametric nonlinear additive time series model of the form

$$Y_t = \mu + \sum_{l=1}^{q} U_{t-l}\beta_l + \sum_{i=1}^{p} \sum_{j=1}^{n_i} f_{ij}(Y_{t-i})\theta_{ij} + e_t. \tag{2.6}$$

To estimate the parameters involved in (2.5), we need to introduce the following symbols. For $1 \leq i \leq p$, let

$$\begin{aligned}
\theta_i &= (\theta_{i1}, \cdots, \theta_{in_i})^\tau, \quad \theta = (\theta_1^\tau, \cdots, \theta_p^\tau)^\tau, \\
F_i &= F_{in_i} = (F_i(V_{1i}), \ldots, F_i(V_{Ti}))^\tau, \quad F = (F_1, F_2, \ldots, F_p), \\
\overline{U} &= \frac{1}{T}\sum_{t=1}^{T} U_t, \quad \widetilde{U} = (U_1 - \overline{U}, \cdots, U_T - \overline{U})^\tau, \\
\overline{Y} &= \frac{1}{T}\sum_{t=1}^{T} Y_t, \quad \widetilde{Y} = (Y_1 - \overline{Y}, \cdots, Y_T - \overline{Y})^\tau, \\
P &= F(F^\tau F)^+ F^\tau, \quad \widehat{U} = (I - P)\widetilde{U}, \quad \widehat{Y} = (I - P)\widetilde{Y}, \tag{2.7}
\end{aligned}$$

and $n = (n_1, \cdots, n_p)^\tau$ and A^+ denotes the Moore–Penrose inverse of A.

Using the approximate model (2.6), we define the least squares (LS) estimators of (β, θ, μ) by

$$\begin{aligned}
\widehat{\beta} &= \widehat{\beta}(n) = \left(\widehat{U}^\tau \widehat{U}\right)^+ \widehat{U}^\tau \widehat{Y}, \\
\widehat{\theta} &= (F^\tau F)^+ F^\tau \left(\widetilde{Y} - \widetilde{U}\widehat{\beta}\right), \\
\widehat{\mu} &= \overline{Y} - \overline{U}^\tau \widehat{\beta}. \tag{2.8}
\end{aligned}$$

Equation (2.8) suggests estimating the conditional mean function $m(X_t) = E[Y_t | X_t]$ by

$$\widehat{m}(X_t; n) = \widehat{\mu} + U_t^\tau \widehat{\beta} + \sum_{i=1}^{p} F_i(V_{ti})^\tau \widehat{\theta}_i(n), \tag{2.9}$$

where $\widehat{\theta}_i(n)$ is the corresponding estimator of θ_i.

It follows from (2.9) that the prediction equation depends on not only the series functions $\{f_{ij} : 1 \leq j \leq n_i, 1 \leq i \leq p\}$ but also n, the vector of truncation parameters. It is mentioned that the choice of the series functions is much less critical than that of the vector of truncation parameters. The series functions used in this chapter need to satisfy Assumptions 2.2 and 2.3 in Section 2.5. The assumptions hold when each f_{ij} belongs to a class of trigonometric series used by Gao, Tong and Wolff (2002a). Therefore, a crucial problem is how to select k practically. Li (1985, 1986, 1987) discussed the asymptotic optimality of a generalized cross–validation (GCV) criterion as well as other model selection criteria. Wahba (1990) provided a recently published survey of nonparametric smoothing spline literature up to 1990. Gao (1998) applied a generalized cross–validation criterion to choose smoothing truncation parameters for the time series case. In this section, we apply a generalized cross–validation method to choose k and then determine the estimates in (2.9).

In order to select n, we introduce the following mean squared error:

$$\widehat{D}(n) = \frac{1}{T} \sum_{t=1}^{T} \{\widehat{m}(X_t; n) - m(X_t)\}^2. \qquad (2.10)$$

Let $g_i^{(m_i)}$ be the m_i–order derivative of the function g_i and M_{0i} be a constant,

$$G_{m_i}(S_i) = \left\{ g : \left| g_i^{(m_i)}(s) - g_i^{(m_i)}(s') \right| \leq M_{0i}|s - s'|, \ s, s' \in S_i \subset \mathbf{R}^1 \right\},$$

where each $m_i \geq 1$ is an integer, $0 < M_{0i} < \infty$ and each S_i is a compact subset of $\mathbf{R}^1$. Let also $\mathcal{N}_{iT} = \{p_{iT}, p_{iT} + 1, \ldots, q_{iT}\}$, in which $p_{iT} = \left[a_i T^{d_i}\right]$, $q_{iT} = \left[b_i T^{c_i}\right]$, $0 < a_i < b_i < \infty$, $0 < d_i < c_i < \frac{1}{2(m_i+1)}$ are constants, and $[x] \leq x$ denotes the largest integer part of x.

Definition 2.1. A data-driven estimator $\widehat{n} = (\widehat{n}_1, \ldots, \widehat{n}_p)^\tau$ is *asymptotically optimal* if

$$\frac{\widehat{D}(\widehat{n})}{\inf_{n \in \mathcal{N}_T} \widehat{D}(n)} \to_p 1,$$

where $n \in \mathcal{N}_T = \{n = (n_1, \ldots, n_p)^\tau : n_i \in \mathcal{N}_{iT}\}$.

Definition 2.2. Select n, denoted by $\widehat{n}_G = (\widehat{n}_{1G}, \ldots, \widehat{n}_{pG})^\tau$, that achieves

$$\mathrm{GCV}(\widehat{n}_G) = \inf_{n \in \mathcal{N}_T} \mathrm{GCV}(n) = \inf_{n \in \mathcal{N}_T} \frac{\widehat{\sigma}^2(n)}{\left[1 - \frac{1}{T} \sum_{i=1}^{p} n_i\right]^2},$$

where $\widehat{\sigma}^2(n) = \frac{1}{T} \sum_{t=1}^{T} \{Y_t - \widehat{m}(X_t; n)\}^2$.

We now have the following asymptotic properties for $\widehat{D}(n)$ and $\widehat{n}_G$.

Theorem 2.1. (i) *Assume that Assumptions 2.1–2.2(i), 2.3 and 2.4 listed in Section 2.5 hold. Then*

$$\widehat{D}(n) = \frac{\sigma^2}{T} \sum_{i=1}^{p} n_i + \frac{1}{T} E\left[\Delta^\tau \Delta\right] + o_p\left(\widehat{D}(n)\right), \qquad (2.11)$$

where $\Delta = \sum_{i=1}^{p} [F_i \theta_i - G_i]$, $G_i = (g_i(V_{1i}), \ldots, g_i(V_{Ti}))^\tau$ *and* $\{F_i(\cdot)\}$ *is as defined before.*

(ii) *In addition, if Assumption 2.2(ii) holds, then we have*

$$\widehat{D}(n) = \frac{\sigma^2}{T} \sum_{i=1}^{p} n_i + \sum_{i=1}^{p} C_i n_i^{-2(m_i+1)} + o_p(\widehat{D}(n)) \qquad (2.12)$$

uniformly over $n \in \mathcal{N}_T$, *where* $\sigma^2 = E[e_t^2] < \infty$ *and each* m_i *is the smoothness order of* g_i.

Theorem 2.2. (i) *Under the conditions of Theorem 2.1(i),* $\widehat{n}_G$ *is asymptotically optimal.*

(ii) *Under the conditions of Theorem 2.1(ii), we have*

$$\frac{\widehat{D}(\widehat{n}_G)}{\widehat{D}(\widehat{n}_D)} - 1 = o_p(T^{-\tau}) \qquad (2.13)$$

and

$$\sum_{i=1}^{p} \left| \frac{\widehat{n}_{iG}}{\widehat{n}_{iD}} - 1 \right| = o_p\left(T^{-\tau}\right), \qquad (2.14)$$

where $\widehat{n}_{iD}$ *is the* i-*th component of* $\widehat{n}_D = (\widehat{n}_{1D}, \ldots, \widehat{n}_{pD})^\tau$ *that minimises* $\widehat{D}(n)$ *over* $\mathcal{N}_T$, $0 < \tau = \min(\tau_1 - \epsilon_1, \tau_2 - \epsilon_2)$, *in which* $\tau_1 = \frac{1}{2} d_{\min}$, $\tau_2 = \frac{1}{2} - 2c_{\max}$, *both* ϵ_1 *and* ϵ_2 *satisfying* $0 < \epsilon_1 < \tau_1$ *and* $0 < \epsilon_2 < \tau_2$ *are arbitrarily small,* $d_{\min} = \min_{1 \leq i \leq p} d_i$ *and* $c_{\max} = \max_{1 \leq i \leq p} c_i$.

The proofs of Theorems 2.1 and 2.2 are relegated to Section 2.5.

We now define the adaptive and simultaneous estimation procedure as follows:

(i) solve the LS estimator $\widehat{\theta}(n)$;

(ii) define the prediction equation by (2.9);

(iii) solve the GCV-based $\widehat{n}_G$; and

(iv) define the following adaptive and simultaneous prediction equation $\widehat{m}(X_t; \widehat{n}_G)$.

If σ^2 is unknown, it will be estimated by $\hat{\sigma}^2(\hat{n}_G)$.

Furthermore, we have the following asymptotic normality.

Corollary 2.1. *Under the conditions of Theorem 2.1(i), we have as* $T \to \infty$

$$\sqrt{T}\left(\hat{\sigma}^2(\hat{n}_G) - \sigma^2\right) \to N\left(0, \mathrm{var}(e_1^2)\right).$$

The proof of Corollary 2.1 is relegated to Section 2.5.

Remark 2.1. Theorem 2.1 provides asymptotic representations for the average squared error $\hat{D}(n)$. See Härdle, Hall and Marron (1988) for an equivalent result in nonparametric kernel regression. In addition, Theorem 2.2(i) shows that the GCV based $\hat{n}_G$ is asymptotically optimal. This conclusion is equivalent to Corollary 3.1 of Li (1987) in the model selection problem. However, the fundamental difference between our discussion in this section and Li (1987) is that we use the GCV method to determine how many terms are required to ensure that each nonparametric function can be approximated optimally, while Li (1987) suggested using the GCV selection criterion to determine how many variables should be employed in a linear model. Due to the different objectives, our conditions and conclusions are different from those of Li (1987), although there are some similarities.

Remark 2.2. Theorem 2.2(ii) not only establishes the asymptotic optimality but also provides the rate of convergence. This rate of convergence is equivalent to that of bandwidth estimates in nonparametric kernel regression. See Härdle, Hall and Marron (1992). More recently, Hurvich and Tsai (1995) have established a similar result for a linear model selection. Moreover, it follows from Theorem 2.2(ii) that the rate of convergence depends heavily on d_i and c_i. Let $d_i = \frac{1}{2m_i+3}$ and $c_i = \frac{1}{2m_i+3} + \eta_i$ for arbitrarily small $\eta_i > 0$. Then the rate of convergence will be of order

$$\min\left(\min_{1 \le i \le p}\left(\frac{1}{2(2m_i+3)}\right), \max_{1 \le i \le p}\left(\frac{2m_i-1}{2(2m_i+3)}\right)\right) - \epsilon$$

for some arbitrarily small $\epsilon > 0$. Obviously, if each g_i is continuously differentiable, then the rate of convergence will be close to $\frac{1}{10} - \epsilon$. This is equivalent to Theorem of Hurvich and Tsai (1995). As a result of the Theorem, the rate of convergence can be close to $\frac{1}{2}$. See also Theorem 1 and Remark 2 of Härdle, Hall and Marron (1992).

Remark 2.3. In this chapter, we assume that the data set $\{(Y_t, X_t) : t \ge 1\}$ satisfies model (2.2) and then propose the orthogonal series method to model the data set. In practice, before applying the estimation procedure to model the data, a crucial problem is how to test the additivity. Some related results for additive nonparametric regression have been given by some authors. See, for example, Gao, Tong and Wolff (2002b).

To illustrate the above estimation procedure, we now include two simulated and real examples for a special case of model (2.2) with $\mu = \beta = 0$. Let $V_t = (V_{t1}, V_{t2}, V_{t3})^\tau = (Y_{t-1}, Y_{t-2}, W_t)^\tau$, where $\{W_t\}$ is to be specified below.

Example 2.1: Consider the model given by

$$Y_t = 0.25Y_{t-1} + 0.25\frac{Y_{t-2}}{1 + Y_{t-2}^2} + \frac{1}{8\pi}W_t^2 + e_t, \ t = 3, 4, ..., T, \quad (2.15)$$

where $\{e_t\}$ is uniformly distributed over $(-0.5\pi, 0.5\pi)$, Y_1 and Y_2 are mutually independent and uniformly distributed over $\left[\frac{1}{128}, 2\pi - \frac{1}{128}\right]$, (Y_1, Y_2) is independent of $\{e_t : t \geq 3\}$,

$$W_t = 0.25W_{t-1} - 0.25W_{t-2} + \epsilon_t, \quad (2.16)$$

in which $\{\epsilon_t\}$ is uniformly distributed over $(-0.5\pi, 0.5\pi)$, X_1 and X_2 are mutually independent and uniformly distributed over $\left[\frac{1}{128}, 2\pi - \frac{1}{128}\right]$, and (X_1, X_2) is independent of $\{\epsilon_t : t \geq 3\}$.

First, it follows from Lemma 3.1 of Masry and Tjøstheim (1997) that both the stationarity and the mixing condition are met. See also Chapter 4 of Tong (1990), §2.4 of Tjøstheim (1994) and §2.4 of Doukhan (1995). Thus, Assumption 2.1(i) holds. Second, it follows from (2.15) and (2.16) that Assumption 2.1(ii) holds immediately. Third, let

$$
\begin{aligned}
g_1(x) &= 0.25x, \\
g_2(x) &= 0.25\frac{x}{1 + x^2}, \\
g_3(x) &= \frac{1}{8\pi}x^2.
\end{aligned}
\quad (2.17)
$$

Since $\{g_i : 1 \leq i \leq 3\}$ are continuously differentiable on $\mathbf{R}^1$, there exist three corresponding periodic functions defined on $[0, 2\pi]$ that are continuously differentiable on $[0, 2\pi]$ and coincide with $\{g_i : 1 \leq i \leq 3\}$ correspondingly (see Hong and White 1995, p.1141). Similarly to §3.2 of Eastwood and Gallant (1991), we can show that there exist the following three corresponding trigonometric polynomials

$$g_1^*(x) = \sum_{j=1}^{n_1} \sin(jx)\theta_{1j},$$

$$g_2^*(x) = \sum_{j=1}^{n_2} \sin(jx)\theta_{2j},$$

$$g_3^*(x) = \sum_{j=1}^{n_3} \cos(jx)\theta_{3j} \tag{2.18}$$

such that Assumptions 2.2(i) and 2.2(ii) are satisfied and the same convergence rate can be obtained as in the periodic case. Obviously, it follows from (2.18) that Assumption 2.2(i) holds. Fourth, Assumption 2.3 is satisfied due to (2.18) and the orthogonality of trigonometric series. Finally, Assumption 2.4 holds due to the fact that $\sup_{t \geq 1} |Y_t| \leq 2\pi$.

We now define g_1^*, g_2^* and g_3^* as the corresponding approximations of g_1, g_2 and g_3 with

$$x \in S = \left[\frac{1}{128}, 2\pi - \frac{1}{128}\right] \text{ and } h_i \in \mathcal{N}_{iT} = \left\{[a_i T^{d_i}], \ldots, [b_i T^{c_i}]\right\}, \tag{2.19}$$

in which $i = 1, 2, 3$,

$$d_i = \frac{1}{2m_i + 3} \text{ and } c_i = \frac{1}{2m_i + 3} + \frac{2m_i - 1}{6(2m_i + 3)}.$$

In the following simulation, we consider the case where $a_i = 1$, $b_i = 2$ and $m_i = 1$ for $i = 1, 2, 3$. Let

$$\begin{aligned}
F_1(x) &= (\sin(x), \sin(2x), \ldots, \sin(n_1 x))^\tau, \\
F_2(x) &= (\sin(x), \sin(2x), \ldots, \sin(n_2 x))^\tau, \\
F_3(x) &= (\cos(x), \cos(2x), \ldots, \cos(n_3 x))^\tau.
\end{aligned}$$

For the cases of $T = 102, 252, 402, 502,$ and 752, we then compute $\widehat{D}(n)$, $\widehat{\sigma}^2(n)$, $\mathrm{GCV}(n)$ and the following quantities: for $i = 1, 2, 3$,

$$d_i(\widehat{n}_{iG}, \widehat{n}_{iD}) = \frac{\widehat{n}_{iG}}{\widehat{n}_{iD}} - 1, \quad d_4(\widehat{n}_G, \widehat{n}_D) = \frac{\widehat{D}(\widehat{n}_G)}{\widehat{D}(\widehat{n}_D)} - 1,$$

$$\mathrm{ASE}_i(\widehat{n}_G) = \frac{1}{N} \sum_{n=1}^{N} \left\{ F_{i\widehat{n}_{iG}}(Z_{ni})^\tau \widehat{\theta}_i(\widehat{h}_G) - g_i(Z_{ni}) \right\}^2,$$

$$\mathrm{ASE}_4(\widehat{n}_G) = \frac{1}{N} \sum_{n=1}^{N} \left\{ \sum_{i=1}^{3} \left(F_{i\widehat{n}_{iG}}(Z_{ni})^\tau \widehat{\theta}_{i\widehat{n}_{iG}} - g_i(Z_{ni}) \right) \right\}^2,$$

$$\mathrm{VAR}(\widehat{n}_G) = \left| \widehat{\sigma}^2(\widehat{n}_G) - \sigma^2 \right|,$$

where $N = T - 2$, $\sigma^2 = \frac{\pi^2}{12} = 0.822467$, $\widehat{n}_G = (\widehat{n}_{1G}, \widehat{n}_{2G}, \widehat{n}_{3G})^\tau$, $Z_{n1} = Y_{n+1}$, $Z_{n2} = Y_n$ and $Z_{n3} = W_{n+2}$.

The simulation results below were performed 1000 times and the means are tabulated in Table 2.1 below.

Table 2.1. Simulation Results for Example 2.1

N	100	250	400	500	750
$\mathcal{N}_{iT}$	$\{1,\ldots,5\}$	$\{1,\ldots,6\}$	$\{1,\ldots,6\}$	$\{1,\ldots,6\}$	$\{1,\ldots,7\}$
$d_1(\widehat{n}_{1G},\widehat{n}_{1D})$	0.10485	0.08755	0.09098	0.08143	0.07943
$d_2(\widehat{n}_{2G},\widehat{n}_{2D})$	0.11391	0.07716	0.08478	0.08964	0.07983
$d_3(\widehat{n}_{3G},\widehat{n}_{3D})$	0.09978	0.08155	0.08173	0.08021	0.08371
$d_4(\widehat{n}_{G},\widehat{n}_{D})$	0.32441	0.22844	0.24108	0.22416	0.22084
$\text{ASE}_1(\widehat{n}_G)$	0.03537	0.01755	0.01123	0.00782	0.00612
$\text{ASE}_2(\widehat{n}_G)$	0.02543	0.01431	0.00861	0.00609	0.00465
$\text{ASE}_3(\widehat{n}_G)$	0.02507	0.01348	0.00795	0.00577	0.00449
$\text{ASE}_4(\widehat{n}_G)$	0.06067	0.03472	0.02131	0.01559	0.01214
$\text{VAR}(\widehat{n}_G)$	0.05201	0.03361	0.01979	0.01322	0.01086

Remark 2.4. Both Theorem 2.2(ii) and Table 2.1 demonstrate that the rate of convergence of the GCV based d_i for $1 \leq i \leq 4$ is of order $T^{-\frac{1}{10}}$. In addition, the simulation results for $\text{ASE}_i(\widehat{n}_G)$ given in Table 2.1 show that when n_i is of order $T^{\frac{1}{5}}$, the rate of convergence of each ASE_i is of order $T^{-\frac{4}{5}}$.

Example 2.2: In this example, we consider the Canadian lynx data. This data set is the annual record of the number of Canadian lynx trapped in the MacKenzie River district of North–West Canada for the years 1821 to 1934. Tong (1976) fitted an eleventh-order linear Gaussian autoregressive model to $Y_t = \log_{10}\{\text{number of lynx trapped in the year } (1820 + t)\}$ for $t = 1, 2, \ldots, 114$ ($T = 114$). It follows from the definition of $\{Y_t, 1 \leq t \leq 114\}$ that all the transformed values $\{Y_t : t \geq 1\}$ are bounded.

We apply the above estimation procedure to fit the real data set listed in Example 2.2 by the following third–order additive autoregressive model of the form

$$Y_t = g_1(Y_{t-1}) + g_2(Y_{t-2}) + g_3(Y_{t-3}) + e_t, \ t = 4, 5, \ldots, T, \qquad (2.20)$$

where $\{g_i : i = 1, 2, 3\}$ are unknown functions, and $\{e_t\}$ is a sequence of independent random errors with zero mean and finite variance.

Similarly, we approximate g_1, g_2 and g_3 by

$$g_1^*(u) = \sum_{j=1}^{n_1} f_{1j}(u)\theta_{1j}, \ g_2^*(v) = \sum_{j=1}^{n_2} f_{2j}(v)\theta_{2j}, \ g_3^*(w) = \sum_{j=1}^{n_3} f_{3j}(w)\theta_{3j},$$

(2.21)

respectively, where $f_{1j}(u) = \sin(ju)$ for $1 \leq j \leq n_1$, $f_{2j}(v) = \sin(jv)$ for $1 \leq j \leq n_2$, $f_{3j}(w) = \cos(jw)$ for $1 \leq j \leq n_3$, and

$$h_j \in \mathcal{N}_{jT} = \left\{ \left[T^{0.2} \right], \ldots, \left[2T^{\frac{7}{30}} \right] \right\}.$$

Our simulation suggests using the following polynomial prediction

$$\widehat{Y}_t = \sum_{j=1}^{\widehat{n}_{1G}} \sin(jY_{t-1})\theta_{1j} + \sum_{j=1}^{\widehat{n}_{2G}} \sin(jY_{t-2})\theta_{2j} + \sum_{j=1}^{\widehat{n}_{3G}} \cos(jY_{t-3})\theta_{3j}, \quad (2.22)$$

where $\widehat{n}_{1G} = 5$, $\widehat{n}_{2G} = \widehat{n}_{3G} = 6$, and the coefficients are given in the following Table 2.2.

Table 2.2. Coefficients for Equation (2.22)

$\theta_1 = (\theta_{11}, \ldots, \theta_{15})^\tau$	$\theta_2 = (\theta_{21}, \ldots, \theta_{26})^\tau$	$\theta_3 = (\theta_{31}, \ldots, \theta_{36})^\tau$
11.877	-2.9211	-6.8698
18.015	-5.4998	-7.8529
10.807	-4.9084	-7.1952
4.1541	-3.1189	-4.8019
0.7997	-1.2744	-2.0529
	-0.2838	-0.4392

The estimator of the error variance was 0.0418. Some plots for Example 2.2 are given in Figure 2.1 of Gao, Tong and Wolff (2002a).

Remark 2.5. For the Canadian lynx data, Tong (1976) fitted an eleventh–order linear Gaussian autoregressive model to the data, and the estimate of the error variance was 0.0437. Figure 2.1 shows that when using equation (2.20) to fit the real data set, the estimator of g_1 is almost linear while the estimators of both g_2 and g_3 appear to be nonlinear. This finding is the same as the conclusion reached by Wong and Kohn (1996), who used a Bayesian based iterative procedure to fit the real data set. Their estimator of the error variance was 0.0421, which is comparable with our variance estimator of 0.0418. Moreover, our estimation procedure provides the explicit equation (2.22) and the CPU time for Example 2.2 just took about 2 minutes. By contrast, Wong and Kohn (1996) can only provide an iterative estimation procedure for each g_i since their approach depends heavily on the Gibbs sampler.

Remark 2.6. Both Examples 2.1 and 2.2 demonstrate that the explicit estimation procedure can not only provide some additional information for further diagnostics and statistical inference but also produce models with better predictive power than is available from linear models. For example, model (2.22) is more appropriate than a completely linear model for the lynx data as mentioned in Remark 2.2. Moreover, model (2.22) not only can be calculated at a new design point with the same convenience as in linear models, but also provides the individual coefficients, which can be used to measure whether the individual influence of each Y_{t-3+i} for $i = 0, 1, 2$ can be negligible.

This section has assumed that the true conditional mean function is of a semiparametric additive model of the form (2.2) and then developed the orthogonal series based estimation procedure. As discussed in the next section, we may approximate the true conditional mean function by the additive form (2.2) even if the true conditional mean function may not be expressed exactly as an additive form.

2.3 Semiparametric kernel estimation

As mentioned above (2.1), we are approximating the mean function $m(U_t, V_t) = E[Y_t | U_t, V_t]$ by minimizing

$$E\left[Y_t - m_1(U_t, V_t)\right]^2 = E\left[Y_t - \mu - U_t^\tau \beta - g(V_t)\right]^2 \qquad (2.23)$$

over a class of functions of the form $m_1(U_t, V_t) = \mu + U_t^\tau \beta + g(V_t)$ with $E[g(V_t)] = 0$. Such a minimization problem is equivalent to minimizing

$$E\left[Y_t - \mu - U_t^\tau \beta - g(V_t)\right]^2 = E\left[E\left\{(Y_t - \mu - U_t^\tau \beta - g(V_t))^2 \,|\, V_t\right\}\right]$$

over some (μ, β, g). This implies that $g(V_t) = E\left[(Y_t - \mu - U_t^\tau \beta)|V_t\right]$ and $\mu = E[Y_t - U_t^\tau \beta]$ with β being given by

$$\beta = \Sigma^{-1} E\left[(U_t - E[U_t|V_t])(Y_t - E[Y_t|V_t])\right] \qquad (2.24)$$

provided that the inverse $\Sigma^{-1} = \left(E\left[(U_t - E[U_t|V_t])(U_t - E[U_t|V_t])^\tau\right]\right)^{-1}$ exists. This also shows that $m_1(U_t, V_t)$ is identifiable under the assumption of $E[g(V_t)] = 0$.

We now turn to estimation assuming that the data are available for (Y_t, U_t, V_t) for $1 \le t \le T$. Since the definitions of the estimators to be used later are quite involved notationally, we start by outlining the main steps in establishing estimators for μ, β and $g(\cdot)$ in (2.1) and then

$g_l(\cdot), l = 1, 2, \cdots, p$ in (2.2). In the following, we give an outline in three steps.

Step 1: Estimating μ and $g(\cdot)$ assuming β to be known.

For each fixed β, since $\mu = E[Y_t] - E[U_t^\tau \beta] = \mu_Y - \mu_U^\tau \beta$, the parameter μ can be estimated by $\hat{\mu}(\beta) = \overline{Y} - \overline{U}^\tau \beta$, where $\mu_Y = E[Y_t]$, $\mu_U = (\mu_U^{(1)}, \cdots, \mu_U^{(q)})^\tau = E[U_t]$, $\overline{Y} = \frac{1}{T}\sum_{t=1}^{T} Y_t$ and $\overline{U} = \frac{1}{T}\sum_{t=1}^{T} U_t$.

Moreover, the conditional expectation

$$
\begin{aligned}
g(x) &= g(x, \beta) = E\left[(Y_t - \mu - U_t^\tau \beta)|V_t = x\right] \\
&= E\left[(Y_t - E[Y_t]) - (U_t - E[U_t])^\tau \beta)|V_t = x\right] \quad (2.25)
\end{aligned}
$$

can be estimated by standard local linear estimation (Fan and Gijbels 1996) with $\hat{g}_T(x, \beta) = \hat{a}_0(\beta)$ satisfying

$$
(\hat{a}_0(\beta), \hat{a}_1(\beta)) = \arg \min_{(a_0, a_1) \in \boldsymbol{R}^1 \times \boldsymbol{R}^p} \quad (2.26)
$$

$$
\times \sum_{t=1}^{T} \left(\tilde{Y}_t - \tilde{U}_t^\tau \beta - a_0 - a_1^\tau(V_t - x)\right)^2 K_t(x, b),
$$

where $\tilde{Y}_t = Y_t - \overline{Y}$, $\tilde{U}_t = (\tilde{U}_t^{(1)}, \cdots, \tilde{U}_t^{(q)})^\tau = U_t - \overline{U}$ and $K_t(x, b) = \prod_{l=1}^{p} K\left(\frac{V_{tl} - x_l}{b_l}\right)$, with $b = b_T = (b_1, \cdots, b_p)$, $b_l = b_{l,T}$ being a sequence of bandwidths for the l-th covariate variable V_{tl}, tending to zero as T tends to infinity, and $K(\cdot)$ is a bounded kernel function on $\boldsymbol{R}^1$.

Step 2: Marginal integration to obtain $g_1, \cdots, g_p$ of (2.2).

The idea of the marginal integration estimator is best explained if $g(\cdot)$ is itself additive, that is, if

$$
g(V_t) = g(V_{t1}, \cdots, V_{tp}) = \sum_{l=1}^{p} g_l(V_{tl}).
$$

Then, since $E[g_l(V_{tl})] = 0$ for $l = 1, \cdots, p$, for k fixed

$$
g_k(x_k) = E\left[g(V_{t1}, \cdots, x_k, \cdots, V_{tp})\right].
$$

An estimate of g_k is obtained by keeping V_{tk} fixed at x_k and then taking the average over the remaining variables $V_{t1}, \cdots, V_{t(k-1)}, V_{t(k+1)}, \cdots, V_{tp}$. This marginal integration operation can be implemented irrespective of whether or not $g(\cdot)$ is additive. If the additivity does not hold, the marginal integration amounts to a projection on the space of additive functions of $V_{tl}, l = 1, \cdots, p$ taken with respect to the product measure of $V_{tl}, l = 1, \cdots, p$, obtaining the approximation $g_a(x, \beta) = \sum_{l=1}^{p} P_{l,\omega}(V_{tl}, \beta)$, which will be detailed below with β appearing linearly in the expression.

In addition, it has been found convenient to introduce a pair of weight functions $(w_k, w_{(-k)})$ in the estimation of each component, hence the index w in $P_{l,w}$. The details are given in Equations (2.32)–(2.36) below.

Step 3: Estimating β.

The last step consists in estimating β. This is done by weighted least squares, and it is easy since β enters linearly in our expressions. In fact, using the expression of $g(x, \beta)$ in Step 1, we obtain the weighted least squares estimator $\hat{\beta}$ of β in (2.34) below. Finally, this is re–introduced in the expressions for $\hat{\mu}$ and $\hat{P}$ resulting in the estimates in (2.35) and (2.36) below. In the following, steps 1–3 are written correspondingly in more detail.

Step 1: To write our expression for $(\hat{a}_0(\beta), \hat{a}_1(\beta))$ in (2.26), we need to introduce some more notation.

$$\mathcal{X}_t = \mathcal{X}_t(x, b) = \left(\frac{(V_{t1} - x_1)}{b_1}, \cdots, \frac{(V_{tp} - x_p)}{b_p} \right)^\tau,$$

and let $b_\pi = \prod_{l=1}^{p} b_l$. We define for $0 \le l_1, l_2 \le p$,

$$\gamma_{T,l_1 l_2} = (Tb_\pi)^{-1} \sum_{t=1}^{T} (\mathcal{X}_t(x, b))_{l_1} (\mathcal{X}_t(x, b))_{l_2} K_t(x, b), \qquad (2.27)$$

where $(\mathcal{X}_t(x, b))_l = \frac{(V_{tl} - x_l)}{b_l}$ for $1 \le l \le p$. We then let $(\mathcal{X}_t(x, b))_0 \equiv 1$ and define

$$\lambda_{T,l}(\beta) = (Tb_\pi)^{-1} \sum_{t=1}^{T} \left(\tilde{Y}_t - \tilde{U}_t^\tau \beta \right) (\mathcal{X}_t(x, b))_l K_t(x, b) \qquad (2.28)$$

and where, as before, $\tilde{Y}_t = Y_t - \bar{Y}$ and $\tilde{U}_t = U_t - \bar{U}$.

Note that $\lambda_{T,l}(\beta)$ can be decomposed as

$$\lambda_{T,l}(\beta) = \lambda_{T,l}^{(0)} - \sum_{s=1}^{q} \beta_s \lambda_{T,l}^{(s)}, \quad \text{for } l = 0, 1, \cdots, p, \qquad (2.29)$$

in which $\lambda_{T,l}^{(0)} = \lambda_{T,l}^{(0)}(x, b) = (Tb_\pi)^{-1} \sum_{t=1}^{T} \tilde{Y}_t (\mathcal{X}_t(x, b))_l K_t(x, b)$,

$$\lambda_{T,l}^{(s)} = \lambda_{T,l}^{(s)}(x, b) = (Tb_\pi)^{-1} \sum_{t=1}^{T} \tilde{U}_{ts} (\mathcal{X}_t(x, b))_l K_t(x, b), \quad 1 \le s \le q.$$

We can then express the local linear estimates in (2.26) as

$$(\hat{a}_0(\beta), \hat{a}_1(\beta) \odot b)^\tau = \Gamma_T^{-1} \Lambda_T(\beta), \qquad (2.30)$$

where $\odot$ is the operation of the component-wise product, i.e., $a_1 \odot b = (a_{11}b_1, \cdots, a_{1p}b_p)$ for $a_1 = (a_{11}, \cdots, a_{1p})$ and $b = (b_1, \cdots, b_p)$,

$$\Lambda_T(\beta) = \begin{pmatrix} \lambda_{T,0}(\beta) \\ \Lambda_{T,1}(\beta) \end{pmatrix}, \quad \Gamma_T = \begin{pmatrix} \gamma_{T,00} & \Gamma_{T,01} \\ \Gamma_{T,10} & \Gamma_{T,11} \end{pmatrix}, \quad (2.31)$$

where $\Gamma_{T,10} = \Gamma_{T,01}^\tau = (\gamma_{T,01}, \cdots, \gamma_{T,0p})^\tau$ and $\Gamma_{T,11}$ is the $p \times p$ matrix defined by $\gamma_{T,l_1 l_2}$ with $l_1, l_2 = 1, \cdots, p$, in (2.27). Moreover, $\Lambda_{T,1}(\beta) = (\lambda_{T,1}(\beta), \ldots, \lambda_{T,p}(\beta))^\tau$ with $\lambda_{T,l}(\beta)$ as defined in (2.28). Analogously for Λ_T, we may define $\Lambda_T^{(0)}$ and $\Lambda^{(s)}$ in terms of $\lambda^{(0)}$ and $\lambda^{(s)}$. Then taking the first component with $c = (1, 0, \cdots, 0)^\tau \in \mathbf{R}^{1+p}$,

$$
\begin{aligned}
\hat{g}_T(x; \beta) &= c^\tau \Gamma_T^{-1}(x) \Lambda_T(x, \beta) \\
&= c^\tau \Gamma^{-1}(x) \Lambda_T^{(0)}(x) - \sum_{s=1}^q \beta_s c^\tau \Gamma^{-1}(x) \Lambda_T^{(s)}(x) \\
&= H_T^{(0)}(x) - \beta^\tau H_T(x),
\end{aligned}
$$

where $H_T(x) = (H_T^{(1)}(x), \cdots, H_T^{(q)}(x))^\tau$ with $H_T^{(s)}(x) = c^\tau \Gamma_T^{-1}(x) \Lambda^{(s)}(x)$, $1 \leq s \leq q$. Clearly, $H_T^{(s)}(x)$ is the local linear estimator of $H^{(s)}(x) = E\left[\left(U_t^{(s)} - \mu_U^{(s)}\right) | V_t = x\right]$, $1 \leq s \leq q$.

We now define $U_t^{(0)} = Y_t$ and $\mu_U^{(0)} = \mu_Y$ such that $H^{(0)}(x) = E[(U_t^{(0)} - \mu_U^{(0)}) | V_t = x] = E[Y_t - \mu_Y | V_t = x]$ and $H(x) = (H^{(1)}(x), \cdots, H^{(q)}(x))^\tau = E[(U_t - \mu_U) | V_t = x]$. It follows that $g(x, \beta) = H^{(0)}(x) - \beta^\tau H(x)$, which equals $g(x)$ under (2.1) irrespective of whether g itself is additive.

Step 2: Let $w_{(-k)}(\cdot)$ be a weight function defined on $\mathbf{R}^{p-1}$ such that $E\left[w_{(-k)}(V_t^{(-k)})\right] = 1$, and $w_k(x_k) = I_{[-L_k, L_k]}(x_k)$ defined on $\mathbf{R}^1$ for some large $L_k > 0$, with

$$V_t^{(-k)} = (V_{t1}, \cdots, V_{t(k-1)}, V_{t(k+1)}, \cdots, V_{tp}),$$

where $I_A(x)$ is the conventional indicator function. In addition, we take

$$V_t(x_k) = (V_{t1}, \cdots, V_{t(k-1)}, x_k, V_{t(k+1)}, \cdots, V_{tp}).$$

For a given β, consider the marginal projection

$$P_{k,w}(x_k, \beta) = E\left[g(V_t(x_k); \beta) w_{(-k)}\left(V_t^{(-k)}\right)\right] w_k(x_k). \quad (2.32)$$

It is easily seen that if g is additive as in (2.2), then for $-L_k \leq x_k \leq L_k$, $P_{k,w}(x_k, \beta) = g_k(x_k)$ up to a constant since it is assumed that

$E\left[w_{(-k)}(V_t^{(-k)})\right] = 1$. In general, $g_a(x, \beta) = \sum_{l=1}^{p} P_{l,w}(x_l, \beta)$ is an additive marginal projection approximation to $g(x)$ in (2.1) up to a constant in the region $x \in \prod_{l=1}^{p}[-L_l, L_l]$. The quantity $P_{k,w}(x_k, \beta)$ can then be estimated by the locally linear marginal integration estimator

$$
\begin{aligned}
\widehat{P}_{k,w}(x_k, \beta) &= T^{-1} \sum_{t=1}^{T} \hat{g}_T(V_t(x_k); \beta) \, w_{(-k)}\left(V_t^{(-k)}\right) \, w_k(x_k) \\
&= \hat{P}_{k,w}^{(0)}(x_k) - \sum_{s=1}^{q} \beta_s \hat{P}_{k,w}^{(s)}(x_k) = \hat{P}_{k,w}^{(0)}(x_k) - \beta^\tau \hat{P}_{k,w}^{U}(x_k),
\end{aligned}
$$

where $\hat{P}_{k,w}^{U}(x_k) = \left(\hat{P}_{k,w}^{(1)}(x_k), \cdots, \hat{P}_{k,w}^{(q)}(x_k)\right)^\tau$, in which

$$
\hat{P}_{k,w}^{(s)}(x_k) = \frac{1}{T} \sum_{t=1}^{T} H_T^{(s)}(V_t(x_k)) \, w_{(-k)}\left(V_t^{(-k)}\right) \, w_k(x_k)
$$

is the estimator of

$$
P_{k,w}^{(s)}(x_k) = E\left[H^{(s)}(V_t(x_k))w_{(-k)}\left(V_t^{(-k)}\right)\right] \, w_k(x_k)
$$

for $0 \le s \le q$ and $P_{k,w}^{U}(x_k) = \left(P_{k,w}^{(1)}(x_k), \cdots, P_{k,w}^{(q)}(x_k)\right)^\tau$ is estimated by $\hat{P}_{k,w}^{U}(x_k)$.

We add the weight function $w_k(x_k) = I_{[-L_k, \ L_k]}(x_k)$ in the definition of $\hat{P}_{k,w}^{(s)}(x_k)$, since we are interested only in the points of $x_k \in [-L_k, \ L_k]$ for some large L_k. In practice, we may use a sample centered version of $\hat{P}_{k,w}^{(s)}(x_k)$ as the estimator of $P_{k,w}^{(s)}(x_k)$. Clearly, we have

$$
P_{k,w}(x_k, \beta) = P_{k,w}^{(0)}(x_k) - \beta^\tau P_{k,w}^{U}(x_k).
$$

Thus, for every β, $g(x) = g(x, \beta)$ of (2.1) (or rather the approximation $g_a(x, \beta)$ if (2.2) does not hold) can be estimated by

$$
\widehat{g}(x, \beta) = \sum_{l=1}^{p} \widehat{P}_{l,w}(x_l, \beta) = \sum_{l=1}^{p} \hat{P}_{l,w}^{(0)}(x_l) - \beta^\tau \sum_{l=1}^{p} \hat{P}_{l,w}^{U}(x_l). \qquad (2.33)
$$

Step 3: We can finally obtain the least squares estimator of β by

$$
\begin{aligned}
\widehat{\beta} &= \arg\min_{\beta \in \mathbf{R}^q} \sum_{t=1}^{T} \left(\tilde{Y}_t - \tilde{U}_t^\tau \beta - \widehat{g}(V_t, \ \beta)\right)^2 \\
&= \arg\min_{\beta \in \mathbf{R}^q} \sum_{t=1}^{T} \left(\widehat{Y}_t^* - \left(\widehat{U}_t^*\right)^\tau \beta\right)^2, \qquad (2.34)
\end{aligned}
$$

where $\widehat{Y}_t^* = \tilde{Y}_t - \sum_{l=1}^p \hat{P}_{l,w}^{(0)}(V_{tl})$ and $\widehat{U}_t^* = \tilde{U}_t - \sum_{l=1}^p \hat{P}_{l,w}^U(V_{tl})$. Therefore,

$$\widehat{\beta} = \left(\sum_{t=1}^T \widehat{U}_t^* \left(\widehat{U}_t^*\right)^\tau\right)^{-1} \left(\sum_{t=1}^T \widehat{Y}_t^* \widehat{U}_t^*\right) \quad \text{and} \quad \widehat{\mu} = \overline{Y} - \widehat{\beta}^\tau \overline{U}. \quad (2.35)$$

We then insert $\widehat{\beta}$ in $\hat{a}_0(\beta) = \hat{g}_{m,n}(x,\beta)$ to obtain $\hat{a}_0(\widehat{\beta}) = \hat{g}_{m,n}(x,\widehat{\beta})$. In view of this, the locally linear projection estimator of $P_k(x_k)$ can be defined by

$$\widehat{\widehat{P}}_{k,w}(x_k) = \frac{1}{T} \sum_{t=1}^T \hat{g}_T(V_t(x_k); \widehat{\beta}) \, w_{(-k)}\left(V_t^{(-k)}\right) \quad (2.36)$$

and for $x_k \in [-L_k, L_k]$ this would estimate $g_k(x_k)$ up to a constant when (2.2) holds. To ensure $E[g_k(V_{tk})] = 0$, we may rewrite

$$\widehat{g}_k(x_k) = \widehat{\widehat{P}}_{k,w}(x_k) - \widehat{\mu}_P(k)$$

for the estimate of $g_k(x_k)$ in (2.2), where $\widehat{\mu}_P(k) = \frac{1}{T} \sum_{t=1}^T \widehat{\widehat{P}}_{k,w}(V_{tk})$.

For the proposed estimators, $\widehat{\beta}$, and $\widehat{\widehat{P}}_{k,w}(\cdot)$, we establish some asymptotic distributions in Theorems 2.3 and 2.4 below under certain technical conditions. To avoid introducing more mathematical details and symbols before we state the main results, we relegate such conditions and their justifications to Section 2.5 of this chapter.

We can now state the asymptotic properties of the marginal integration estimators for both the parametric and nonparametric components. Let $U_t^* = U_t - \mu_U - \sum_{l=1}^p P_{l,w}^U(V_{tl})$, $Y_t^* = Y_t - \mu_Y - \sum_{l=1}^p P_{l,w}^{(0)}(V_{tl})$ and $R_t = U_t^*(Y_t^* - U_t^{*\tau}\beta)$.

Theorem 2.3. *Assume that Assumptions 2.5–2.9 listed in Section 2.5 hold. Then as $T \to \infty$*

$$\sqrt{T}\left[(\widehat{\beta} - \beta) - \mu_\beta\right] \to_D N(0, \Sigma_\beta) \quad (2.37)$$

with $\mu_\beta = \left(B^{UU}\right)^{-1}\mu_B$ and $\Sigma_\beta = \left(B^{UU}\right)^{-1}\Sigma_B\left(\left(B^{UU}\right)^{-1}\right)^\tau$, where $B^{UU} = E\left[U_1^* U_1^{*\tau}\right]$, $\mu_B = E[R_0]$ and $\Sigma_B = E\left[(R_0 - \mu_B)(R_0 - \mu_B)^\tau\right]$.

Furthermore, when (2.2) holds, we have

$$\mu_\beta = 0 \quad \text{and} \quad \Sigma_\beta = \left(B^{UU}\right)^{-1}\Sigma_B\left(\left(B^{UU}\right)^{-1}\right)^\tau, \quad (2.38)$$

where $\Sigma_B = E\left[R_0 R_0^\tau\right]$ with $R_t = U_t^* \varepsilon_t$, and $\varepsilon_t = Y_t - m_1(U_t, V_t) = Y_t - \mu - U_t^\tau\beta - g(V_t)$.

Remark 2.7. Note that

$$
\sum_{l=1}^{p} P_{l,w}^{(0)}(V_{tl}) - \beta^\tau \sum_{l=1}^{p} P_{l,w}^{U}(V_{tl}) = \sum_{l=1}^{p} \left(P_{l,w}^{(0)}(V_{tl}) - \beta^\tau P_{l,w}^{U}(V_{tl}) \right)
$$
$$
= \sum_{l=1}^{p} P_{l,w}(V_{tl}, \beta) \equiv g_a(V_t, \beta).
$$

Therefore $Y_t^* - U_t^{*\tau}\beta = \varepsilon_t + g(V_t) - g_a(V_t, \beta)$, where $g(V_t) - g_a(V_t, \beta)$ is the residual due to the additive approximation. When (2.2) holds, it means that $g(V_t)$ in (2.1) has the expressions

$$
g(V_t) = \sum_{l=1}^{p} g_l(V_{tl}) = \sum_{l=1}^{p} P_{l,w}(V_{tl}, \beta) = g_a(V_t, \beta)
$$

and $H(V_t) = \sum_{l=1}^{p} P_{l,w}^{U}(V_{tl})$, and hence $Y_t^* - U_t^{*\tau}\beta = \varepsilon_t$. As β minimizes $L(\beta) = E[Y_t - m_1(U_t, V_t)]^2$, we have $L'(\beta) = 0$ and $E[\epsilon_t U_t^*] = E[\epsilon_{ij}(U_t - E[U_t|V_t])] = 0$ when (2.2) holds. This implies $E[R_t] = 0$ and hence $\mu_\beta = 0$ in (2.37) when the marginal integration estimation procedure is employed for the additive form of $g(\cdot)$.

In both theory and practice, we need to test whether $H_0 : \beta = \beta_0$ holds for a given β_0. The case where $\beta_0 \equiv 0$ is an important one. Before we state the next theorem, some additional notation is needed. Let

$$
\widehat{B}^{UU} = \frac{1}{T} \sum_{t=1}^{T} \widehat{U}_t^*(\widehat{U}_t^*)^\tau, \quad \widehat{Z}_t^* = \tilde{Z}_t - \sum_{l=1}^{p} \widehat{P}_{l,w}^{U}(V_{tl}),
$$
$$
\widehat{\mu}_B = \frac{1}{T} \sum_{t=1}^{T} \widehat{R}_t, \quad \widehat{R}_t = \widehat{U}_t^* \left(\widehat{Y}_t^* - \left(\widehat{U}_t^*\right)^\tau \widehat{\beta} \right),
$$
$$
\widehat{\mu}_\beta = \left(\widehat{B}^{UU}\right)^{-1} \widehat{\mu}_B, \quad \widehat{\Sigma}_\beta = \left(\widehat{B}^{UU}\right)^{-1} \widehat{\Sigma}_B \left(\left(\widehat{B}^{UU}\right)^{-1} \right)^\tau,
$$

in which $\widehat{\Sigma}_B$ is a consistent estimator of Σ_B, defined simply by

$$
\widehat{\Sigma}_B = \begin{cases} \frac{1}{T} \sum_{t=1}^{T} (\widehat{R}_t - \widehat{\mu}_B)(\widehat{R}_t - \widehat{\mu}_B)^\tau & \text{if (2.1) holds}, \\ \frac{1}{T} \sum_{t=1}^{T} \widehat{R}_t \widehat{R}_t^\tau & \text{if (2.2) holds}. \end{cases}
$$

It can be shown that both $\widehat{\mu}_\beta$ and $\widehat{\Sigma}_\beta$ are consistent estimators of μ_β and Σ_β, respectively.

We now state a corollary of Theorem 2.3 to test hypotheses about β.

Corollary 2.2. *Assume that the conditions of Theorem 2.3 hold. Then* as $T \to \infty$

$$\widehat{\Sigma}_\beta^{-1/2} \sqrt{T} \left[(\widehat{\beta} - \beta) - \widehat{\mu}_\beta \right] \to_D N(0, I_q),$$

$$T \left[(\widehat{\beta} - \beta) - \widehat{\mu}_\beta \right]^\tau \widehat{\Sigma}_\beta^{-1} \left[(\widehat{\beta} - \beta) - \widehat{\mu}_\beta \right] \to_D \chi_q^2.$$

Furthermore, when (2.2) holds, we have as $T \to \infty$,

$$\widehat{\Sigma}_\beta^{-1/2} \sqrt{T} \left(\widehat{\beta} - \beta \right) \to_D N(0, I_q),$$

$$\left(\sqrt{T}(\widehat{\beta} - \beta) \right)^\tau \widehat{\Sigma}_\beta^{-1} \left(\sqrt{T}(\widehat{\beta} - \beta) \right) \to_D \chi_q^2.$$

The proof of Theorem 2.3 is relegated to Section 2.5 while the proof of Corollary 2.2 is straightforward and therefore omitted.

Next we state the following theorem for the nonparametric component.

Theorem 2.4. *Assume that Assumptions 2.5–2.9 listed in Section 2.5 hold. Then for* $x_k \in [-L_k, L_k]$,

$$\sqrt{T\, b_k} \left(\widehat{P}_{k,w}(x_k) - P_{k,w}(x_k) - \text{bias}_{1k} \right) \to_D N(0, \text{var}_{1k}), \qquad (2.39)$$

where

$$\text{bias}_{1k} = \frac{1}{2} b_k^2\, \mu_2(K) \int w_{(-k)}(x^{(-k)}) f_{(-k)}(x^{(-k)}) \frac{\partial^2 g(x, \beta)}{\partial x_k^2}\, dx^{(-k)}$$

and

$$\text{var}_{1k} = J \int V(x, \beta) \frac{[w_{(-k)}(x^{(-k)}) f_{(-k)}(x^{(-k)})]^2}{f(x)}\, dx^{(-k)}$$

with $J = \int K^2(u)du$, $\mu_2(K) = \int u^2 K(u)du$,

$$g(x, \beta) = E\left[(Y_{ij} - \mu - Z_{ij}^\tau \beta)\, | X_{ij} = x \right],$$

and $V(x, \beta) = E\left[(Y_{ij} - \mu - Z_{ij}^\tau \beta - g(x, \beta))^2 | X_{ij} = x \right]$.

Furthermore, assume that (2.2) holds and that $E\left[w_{(-k)}(X_{ij}^{(-k)}) \right] = 1$. *Then as* $T \to \infty$

$$\sqrt{T\, b_k} \left(\widehat{g}_k(x_k) - g_k(x_k) - \text{bias}_{2k} \right) \to_D N(0, \text{var}_{2k}), \qquad (2.40)$$

where

$$\text{bias}_{2k} = \frac{1}{2} b_k^2\, \mu_2(K) \frac{\partial^2 g_k(x_k)}{\partial x_k^2},$$

$$\text{var}_{2k} = J \int V(x, \beta) \frac{[w_{(-k)}(x^{(-k)}) f_{(-k)}(x^{(-k)})]^2}{f(x)}\, dx^{(-k)}$$

with $V(x, \beta) = E\left[\left(Y_{ij} - \mu - Z_{ij}^\tau \beta - \sum_{k=1}^p g_k(x_k)\right)^2 | X_{ij} = x\right]$.

The proof of Theorem 2.4 is relegated to Section 2.5. Theorems 2.3 and 2.4 may be applied to estimate various additive models such as model (2.2). In the following example, we apply the proposed estimation procedure to determine whether a partially linear time series model is more appropriate than either a completely linear time series model or a purely nonparametric time series model for a given set of real data.

Example 2.3: In this example, we continue analyzing the Canadian lynx data with $y_t = \log_{10}\{$number of lynx trapped in the year $(1820 + t)\}$ for $t = 1, 2, ..., 114$ $(T = 114)$. Let $q = p = 1$, $U_t = y_{t-1}$, $V_t = y_{t-2}$ and $Y_t = y_t$ in model (2.1). We then select y_t as the present observation and both y_{t-1} and y_{t-2} as the candidates of the regressors.

Model (2.1) reduces to a partially linear time series model of the form

$$y_t = \beta y_{t-1} + g(y_{t-2}) + e_t. \tag{2.41}$$

In addition to estimating β and $g(\cdot)$, we also propose to choose a suitable bandwidth h based on a nonparametric cross–validation (CV) selection criterion. For $i = 1, 2$, define

$$\begin{aligned}
\widehat{g}_{i,t}(\cdot) &= \widehat{g}_{i,t}(\cdot, h) \\
&= \frac{1}{T-3} \frac{\sum_{s=3, s \neq t}^T K\left(\frac{\cdot - y_{s-2}}{h}\right) y_{s+1-i}}{\widehat{\pi}_{h,t}(\cdot)}, \tag{2.42}
\end{aligned}$$

where $\widehat{\pi}_{h,t}(\cdot) = \frac{1}{T-3} \sum_{s=3, s \neq t}^T K\left(\frac{\cdot - y_{s-2}}{h}\right)$.

We now define a new LS estimate $\widetilde{\beta}(h)$ of β by minimizing

$$\sum_{t=1}^{T-3} \{y_t - \beta y_{t-1} - \widehat{g}_{1,t}(y_{t-2}) - \beta \widehat{g}_{2,t}(y_{t-2})\}^2.$$

The CV selection function is then defined by

$$CV(h) = \frac{1}{T-3} \sum_{t=3}^T \left\{y_t - \left[\widetilde{\beta}(h) y_{t-1} + \widehat{g}_{1,t}(y_{t-2}) - \widetilde{\beta}(h)\widehat{g}_{2,t}(y_{t-2})\right]\right\}^2. \tag{2.43}$$

For Example 2.3, we choose

$$K(x) = \frac{1}{\sqrt{2\pi}} e^{-\frac{x^2}{2}} \text{ and } H_{114} = [0.3 \cdot 114^{-\frac{7}{30}}, 1.1 \cdot 114^{-\frac{1}{6}}].$$

Before selecting the bandwidth interval H_{114}, we actually calculated the following CV function $CV(h)$ over all possible intervals. Our computation indicates that H_{114} is the smallest possible interval, on which $CV(h)$ can attain their smallest value. Similarly, we conducted a simulation study for the case where K is a uniform kernel before choosing the standard normal kernel. Our simulation results also show that for the lynx data, nonparametric normal kernel estimation procedures can provide more stable simulation results. Example 3 of Yao and Tong (1994) also suggests using the standard normal kernel in the nonparametric fitting of the lynx data.

Through minimising the CV functions $CV(h)$, we obtain $\text{CV}(\widehat{h}_C) = \inf_{h \in H_{114}} CV(h) = 0.04682$. The estimate of the error variance of e_n was 0.04119. In comparison, the estimate of the error variance of the model of Tong (1976) was 0.0437, while the estimate of the error variance of the model of Wong and Kohn (1996) was 0.0421, which is comparable with our variance estimate of 0.04119. Some plots for Example 2.3 are given in Figure 2.1 of Gao and Yee (2000).

For the lynx data set, when selecting y_{t-1} and y_{t-2} as the candidates of the regressors, our research suggests using the prediction equation

$$\widehat{y}_t = 1.354\, y_{t-1} + \widetilde{g}_1(y_{t-2}), \ t = 3, 4, \ldots, \tag{2.44}$$

where

$$\widetilde{g}_1(y_{t-2}) = \widehat{g}_1(y_{t-2}, \widehat{h}_C) - 1.354\, \widehat{g}_2(y_{t-2}, \widehat{h}_C)$$

and

$$\widehat{g}_i(y_{t-2}, h) = \frac{\sum_{s=3}^{T} K\left(\frac{y_{t-2} - y_{s-2}}{h}\right) y_{s+1-i}}{\sum_{u=3}^{T} K\left(\frac{y_{t-2} - y_{u-2}}{h}\right)},$$

in which $i = 1, 2$ and $\widehat{h}_C = 0.1266$. The research by Gao and Yee (2000) clearly shows that $\widetilde{g}_1(\cdot)$ appears to be nonlinear.

2.4 Semiparametric single–index estimation

Consider a semiparametric single–index model of the form

$$Y_t = X_t^\tau \theta + \psi(X_t^\tau \eta) + e_t, \ t = 1, 2, \cdots, T, \tag{2.45}$$

where $\{X_t\}$ is a strictly stationary time series, both θ and η are vectors of unknown parameters, $\psi(\cdot)$ is an unknown function defined over $\mathbf{R}^1$, and $\{e_t\}$ is a sequence of strictly stationary time series errors with $E[e_t|X_t] = 0$ and $E[e_t^2|X_t = x] = \sigma^2(x)$ when our interest is on the estimation of the conditional mean function. Throughout this chapter, we assume that there is some positive constant parameter $\sigma > 0$ such that

$P\left(\sigma^2(X_t) = \sigma^2\right) = 1$. As pointed out in Xia, Tong and Li (1999), model (2.45) covers various special and important cases already discussed extensively in the literature.

In order to estimate the unknown parameters and function involved in (2.45), we introduce the following notation:

$$
\begin{aligned}
\psi_{1\eta}(u) &= E\left[Y_t | X_t^\tau \eta = u\right], \quad \psi_{2\eta}(u) = E\left[X_t | X_t^\tau \eta = u\right], \\
W(\eta) &= E\left[(X_t - \psi_{2\eta}(X_t^\tau \eta))(X_t - \psi_{2\eta}(X_t^\tau \eta))^\tau\right], \\
V(\eta) &= E\left[(X_t - \psi_{2\eta}(X_t^\tau \eta))(Y_t - \psi_{1\eta}(X_t^\tau \eta))^\tau\right], \\
S(\theta, \eta) &= E\left[Y_t - \psi_{1\eta}(X_t^\tau \eta) - \theta^\tau(X_t - \psi_{2\eta}(X_t^\tau \eta))\right]. \quad (2.46)
\end{aligned}
$$

The following theorem shows that model (2.45) is identifiable under some mild conditions; its proof is the same as that of Theorem 2 of Xia, Tong and Li (1999).

Theorem 2.5. *Assume that $\psi(\cdot)$ is twice differentiable and that the marginal density function of $\{X_t\}$ is a positive function on an open convex subset in $\mathbf{R}^d$. Then the minimum point of $S(\theta, \eta)$ with $\theta \perp \eta$ is unique at η and $\theta = \theta_\eta = W^+(\eta)V(\eta)$, where $\theta \perp \eta$ denotes that θ and η are orthogonal and $W^+(\eta)$ is the Moore–Penrose inverse.*

We now start to estimate the identifiable parameters and function. Let $\mathcal{X} \subset \mathbf{R}^1$ be the union of a number of open convex sets such that the marginal density, $\pi(\cdot)$, of $\{X_t\}$ is greater than $M > 0$ on $\mathcal{X}$ for some constant $M > 0$. We first estimate $\psi_{1\eta}(\cdot)$ and $\psi_{2\eta}(\cdot)$ by

$$
\begin{aligned}
\widehat{\psi}_{1\eta}(u) &= \frac{\sum_{X_t \in \mathcal{X}} \widetilde{K}_h(X_t^\tau \eta - u) Y_t}{\sum_{X_t \in \mathcal{X}} \widetilde{K}_h(X_t^\tau \eta - u)}, \\
\widehat{\psi}_{2\eta}(u) &= \frac{\sum_{X_t \in \mathcal{X}} \widetilde{K}_h(X_t^\tau \eta - u) X_t}{\sum_{X_t \in \mathcal{X}} \widetilde{K}_h(X_t^\tau \eta - u)}, \quad (2.47)
\end{aligned}
$$

where $\widetilde{K}_h(\cdot) = \widetilde{K}(\cdot/h)$, $\widetilde{K}(\cdot)$ is a kernel function, and h is a bandwidth parameter. Let

$$
\begin{aligned}
\widetilde{Y}_{t\eta} &= Y_t - \widehat{\psi}_{1\eta}(X_t^\tau \eta), \quad \widetilde{X}_{t\eta} = X_t - \widehat{\psi}_{2\eta}(X_t^\tau \eta), \\
S_T(\theta, \eta; h) &= \sum_{X_t \in \mathcal{X}} \left(\widetilde{Y}_{t\eta} - \widetilde{X}_{t\eta}^\tau \theta\right)^2. \quad (2.48)
\end{aligned}
$$

By minimising $S_T(\theta, \eta; h)$ over (θ, η, h), we may find the least–squares type of estimators for the parameters. First, given (η, h), the corresponding estimator of θ is given by

$$\widehat{\theta}(\eta, h) = \left(\sum_{X_t \in \mathcal{X}} \widetilde{X}_{t\eta} \widetilde{X}_{t\eta}^\tau \right)^+ \sum_{X_t \in \mathcal{X}} \widetilde{X}_{t\eta} \widetilde{Y}_{t\eta}. \qquad (2.49)$$

We then estimate (η, h) by $(\widehat{\eta}, \widehat{h})$ through minimising

$$\widehat{S}_T(\eta, h) = \sum_{X_t \in \mathcal{X}} \left(\widetilde{Y}_{t\eta} - \widetilde{X}_{t\eta}^\tau \widehat{\theta}(\eta, h) \right)^2. \qquad (2.50)$$

The nonparametric estimator of $\psi(\cdot)$ is finally defined by

$$\widehat{\psi}(u) = \widehat{\psi}_{1\widehat{\eta}}(u) - \widehat{\psi}_{1\widehat{\eta}}(u)^\tau \widehat{\theta}(\widehat{\eta}, \widehat{h}). \qquad (2.51)$$

When $\sigma^2 = E[e_t^2]$ is unknown, it is estimated by

$$\widehat{\sigma}^2 = \frac{1}{\widetilde{T}} \, \widehat{S}_T(\widehat{\eta}, \widehat{h}) = \frac{1}{\widetilde{T}} \sum_{X_t \in \mathcal{X}} \left(\widetilde{Y}_{t\widehat{\eta}} - \widetilde{X}_{t\widehat{\eta}}^\tau \widehat{\theta}(\widehat{\eta}, \widehat{h}) \right)^2, \qquad (2.52)$$

where $\widetilde{T} = \# \{ X_t : \ X_t \in \mathcal{X} \}$.

To establish the main theorem of this section, we need to introduce the following notation:

$$A_{1T}(\eta) = \sum_{X_t \in \mathcal{X}} \widetilde{X}_{t\eta} \psi'(X_t^\tau \eta) e_t, \quad A_{2T}(\eta) = \sum_{X_t \in \mathcal{X}} \widetilde{X}_{t\eta} e_t,$$

$$B_1(\eta) = E\left[\widetilde{X}_{t\eta} \widetilde{X}_{t\eta}^\tau \right], \quad B_2(\eta) = E\left[\widetilde{X}_{t\eta} \widetilde{X}_{t\eta}^\tau \psi'(X_t^\tau \eta) \right],$$

$$B_3(\eta) = E\left[\widetilde{X}_{t\eta} \widetilde{X}_{t\eta}^\tau \left(\psi'(X_t^\tau \eta) \right)^2 \right],$$

$$C_1(\eta) = \frac{1}{\widetilde{T}} \mathrm{var}\left[A_{2T}(\eta) - B_2(\eta) B_3^+(\eta) A_{1T}(\eta) \right],$$

$$C_2(\eta) = \frac{1}{\widetilde{T}} \mathrm{var}\left[A_{1T}(\eta) - B_1(\eta) B_3^+(\eta) A_{2T}(\eta) \right], \qquad (2.53)$$

and $C_3 = E\left[\left(e_t^2 - \sigma^2 \right)^2 \right]$.

Theorem 2.6. *Assume that Assumption 2.10 listed in Section 2.5 holds. Then as* $\widetilde{T} \to \infty$

$$\sqrt{\widetilde{T}} \left(\widetilde{\theta} - \theta \right) \to_D N\left(0, C_1^+(\eta) \right),$$

$$\sqrt{\widetilde{T}} \left(\widetilde{\eta} - \eta \right) \to_D N\left(0, C_2^+(\eta) \right),$$

$$\sqrt{\widetilde{T}} \left(\widehat{\sigma}^2 - \sigma^2 \right) \to_D N\left(0, C_3 \right), \quad \frac{\widehat{h}}{h_0} \to_P 1, \qquad (2.54)$$

where h_0 is the theoretically optimal bandwidth in the sense that it min-imizes

$$\text{MISE}(h) = \int E\left[\widehat{\psi}(x^\tau \eta) - \psi(x^\tau \eta)\right]^2 \pi(x)dx.$$

In addition,

$$\sup_{x \in \mathcal{X}} \left|\widehat{\psi}\left(x^\tau \widehat{\eta}\right) - \psi\left(x^\tau \widehat{\eta}\right)\right| = O\left(\sqrt{T^{-\frac{4}{5}}\log(T)}\right) \quad \text{almost surely.}$$

Before the proof of Theorem 2.6 is given in Section 2.5, we examine the finite–sample performance of Theorem 2.6 in Example 2.4 below.

Example 2.4: Consider a nonlinear time series model of the form

$$y_t = 0.3x_t + 0.4x_{t-1} + e^{-2(0.8x_t - 0.6x_{t-1})^2} + 0.1\epsilon_t \qquad (2.55)$$

with $x_t = 0.8x_{t-1} + \xi_t + 0.5\xi_{t-1}$, where $\{\epsilon_t\}$ and $\{\xi_t\}$ are mutually independent random errors drawn from $N(0,1)$. Model (2.55) may be written as

$$y_t = \lambda \cos(\zeta)x_t + \lambda \sin(\zeta)x_{t-1} + e^{-2(\sin(\zeta)x_t - \cos(\zeta)x_{t-1})^2} + 0.1\epsilon_t \quad (2.56)$$

with $\alpha = 0.6435$ and $\lambda = 0.5$.

For sample sizes $n = 50$, 100 and 200, the simulation was replicated 500 times. We choose $\mathcal{X}$ such that $(0.8, -0.6)x \in [-1.5, 1.5]$, where x is a vector, and use the Epanechnikov kernel. We minimize $\widehat{S}_T(\eta, h)$ within $\alpha \in [0.2, 1.3]$, $\eta = (\cos(\zeta), -\sin(\zeta))^\tau$ and $h \in [0.01, 0.2]$. Table 2.3 provides the estimates of $\widehat{\zeta}$, $\widehat{\theta}$ and $\widehat{\eta}$.

Table 2.3. Means and Standard Deviations, in parentheses, of
Estimators for Different Sample Sizes T

T	$\widehat{\zeta}$	$\widehat{\theta}$
50	0.6414 (0.0338)	0.2993 (0.0058)
		0.4008 (0.0084)
100	0.6442 (0.0076)	0.3002 (0.0041)
		3.9970 (0.0042)
200	0.6436 (0.0040)	0.2999 (0.0027)
		0.4000 (0.0033)

Table 2.3 is taken from the first part of Table 1 of the paper by Xia, Tong and Li (1999). Some other results have also been included in the paper.

2.5 Technical notes

Before we complete the proofs of the theorems, we first need to introduce
the following assumptions and definitions.

2.5.1 Assumptions

Assumption 2.1. (i) Assume that the process $\{(X_t, Y_t) : 1 \leq t \leq T\}$
is strictly stationary and α–mixing with the mixing coefficient $\alpha(T) \leq$
$C_\alpha \eta^T$, where $0 < C_\alpha < \infty$ and $0 < \eta < 1$ are constants.

(ii) Let $e_t = Y_t - E[Y_t | X_t]$. Assume that $\{e_t\}$ satisfies for all $t \geq 1$,

$$E[e_t | \Omega_{t-1}] = 0, \quad E[e_t^2 | \Omega_{t-1}] = E[e_t^2] \quad \text{and} \quad E[e_t^4 | \Omega_{t-1}] < \infty$$

almost surely, where $\Omega_t = \sigma\{(X_{s+1}, Y_s) : 1 \leq s \leq t\}$ is a sequence of
σ–fields generated by $\{(X_{s+1}, Y_s) : 1 \leq s \leq t\}$.

Assumption 2.2. (i) For $g_i \in G_{m_i}(S_i)$ and $\{f_{ij}(\cdot) : j = 1, 2, \ldots\}$ given
above, there exists a vector of unknown parameters $\theta_i = (\theta_{i1}, \ldots, \theta_{in_i})^\tau$
such that for a sequence of constants $\{B_i : 1 \leq i \leq p\}$ $(0 < B_i < \infty$
independent of T) and large enough T

$$\sup_{x_i \in S_i} |F_i(x_i)^\tau \theta_i - g_i(x_i)| \leq B_i n_i^{-(m_i+1)} \tag{2.57}$$

uniformly over $n_i \in \mathcal{N}_{iT}$ and $1 \leq i \leq p$, where $\mathcal{N}_{iT} = \{p_{iT}, p_{iT} +$
$1, \ldots, q_{iT}\}$, in which $p_{iT} = [a_i T^{d_i}]$, $q_{iT} = [b_i T^{c_i}]$, $0 < a_i < b_i < \infty$,
$0 < d_i < c_i < \frac{1}{2(m_i+1)}$ are constants, and $[x] \leq x$ denotes the largest
integer part of x.

(ii) Furthermore, there exists a sequence of constants $\{C_i : 1 \leq i \leq p\}$
$(0 < C_i < \infty$ independent of T) such that for large enough T

$$n_i^{2(m_i+1)} E\left\{F_i(V_{ti})^\tau \theta_i - g_i(V_{ti})\right\}^2 \approx C_i \tag{2.58}$$

uniformly over $h_i \in \mathcal{N}_{iT}$ and $1 \leq i \leq p$, where the symbol " $\approx$ " indicates
that the ratio of the left–hand side and the right–hand side tends to one
as $T \to \infty$.

Assumption 2.3. (i) $\{F_i\}$ is of full column rank $n_i \in \mathcal{N}_{iT}$ as T large
enough, $\{f_{ij} : 1 \leq j \leq n_i, 1 \leq i \leq p\}$ are continuous functions with
$\sup_x \sup_{i,j \geq 1} |f_{ij}(x)| \leq c_0 < \infty$.

(ii) Assume that for all $1 \leq i, j \leq p$ and $s \neq t$

$$E[f_{ij}(X_{si}) f_{ij}(X_{ti})] = 0$$

and for all $t \geq 1$

$$E\left[f_{ij}(X_{ti})f_{lm}(X_{tl})\right] = \begin{cases} c_{ij}^2 & \text{if } i = l \text{ and } j = m \\ 0 & \text{if } (i, j, l, m) \in IJLM, \end{cases}$$

where $IJLM = \{(i, j, l, m) : 1 \leq i, l \leq p, 1 \leq j \leq n_i, 1 \leq m \leq n_l\} - \{(i, j, l, m) : 1 \leq i = l \leq p, 1 \leq j = m \leq n_i\}$.

Assumption 2.4. There are positive constants $\{C_i : i \geq 1\}$ such that for $i = 1, 2, \ldots$

$$\sup_x E(|Y_t|^i | X_t = x) \leq C_i < \infty.$$

Assumption 2.5. Assume that the joint probability density function $\pi_s(v_1, \cdots, v_s)$ of $(V_{t_1}, \cdots, V_{t_s})$ exists and is bounded for $s = 1, \cdots, 2r - 1$, where r is some positive integer such that Assumption 2.6(ii) below holds. For $s = 1$, we write $\pi(v)$ for $\pi_1(v_1)$, the marginal density function of V_t.

Assumption 2.6. (i) Let $U_t^* = U_t - \mu_U - \sum_{l=1}^p P_{l,w}^U(V_{tl})$ and $B^{UU} = E[U_1^*(U_1^*)^\tau]$. The inverse matrix of B^{UU} exists. Let $Y_t^* = Y_t - \mu_Y - \sum_{l=1}^p P_{l,w}^{(0)}(V_{tl})$ and $R_t = U_t^*(Y_t^* - U_t^{*\tau}\beta)$. Assume that matrix $\Sigma_B = E[(R_1 - \mu_B)(R_1 - \mu_B)^\tau]$ is finite.

(ii) Suppose that there is some $\lambda > 2$ such that $E\left[|Y_t|^{\lambda r}\right] < \infty$ for r as defined in Assumption 2.5.

Assumption 2.7. (i) The functions $g(\cdot)$ in (2.1) and $g_l(\cdot)$ for $1 \leq l \leq p$ in (2.2) have bounded and continuous derivatives up to order 2. In addition, the function $g(\cdot)$ has a second–order derivative matrix $g''(\cdot)$ (of dimension $p \times p$), which is uniformly continuous on $\mathbf{R}^p$.

(ii) For each k, $1 \leq k \leq p$, the weight function $\{w_{(-k)}(\cdot)\}$ is uniformly continuous on $\mathbf{R}^{p-1}$ and bounded on the compact support $S_w^{(-k)}$ of $w_{(-k)}(\cdot)$. In addition, $E\left[w_{(-k)}\left(X_{ij}^{(-k)}\right)\right] = 1$. Let $S_W = S_{W,k} = S_w^{(-k)} \times [-L_k, L_k]$ be the compact support of $W(x) = W(x^{(-k)}, x_k) = w_{(-k)}\left(x^{(-k)}\right) \cdot I_{[-L_k, L_k]}(x_k)$. In addition, let $\inf_{x \in S_W} \pi(x) > 0$ hold.

Assumption 2.8. $K(x)$ is a symmetric and bounded probability density function on $\mathbf{R}^1$ with compact support, C_K, and finite variance such that $|K(x) - K(y)| \leq M|x - y|$ for x, $y \in C_K$ and $0 < M < \infty$.

Assumption 2.9. (i) Let b_π be as defined before. The bandwidths satisfy

$$\lim_{T \to \infty} \max_{1 \leq l \leq p} b_l = 0, \quad \lim_{T \to \infty} T b_\pi^{1+2/r} = \infty \text{ and } \lim_{T \to \infty} \inf T b_\pi^{\frac{2(r-1)a+2(\lambda r-2)}{(a+2)\lambda}} > 0$$

for some integer $r \geq 3$ and some $\lambda > 2$ being the same as in Assumptions 2.5 and 2.6.

Assumption 2.10. (i) The functions $\psi_\eta(u)$, $\psi_{1\eta}(u)$, $\psi_{2\eta}(u)$ and $\pi_\eta(u)$ have bounded, continuous second derivatives on $\mathcal{U} = \{u = x^\tau \eta : x \in \mathbf{R}^d\}$, where $\pi_\eta(u)$ is the marginal density function of $u = x^\tau \eta$.

(ii) There is some constant $c_\pi > 1$ such that $c_\pi^{-1} < \pi(x) < c_\pi$ for all $x \in \mathcal{X}$. In addition, $\pi(x)$ has bounded second derivative on $x \in \mathcal{X}$.

(iii) For each given η, the conditional density functions $f_{X_1^\tau \tau | v}(u, v)$ and $f_{(X_1^\tau \tau, X_l^\tau \eta) | (v_1, v_2)}(u_1, u_l | v_1, v_l)$ are bounded for all $l > 1$.

(iv) $\widetilde{K}(\cdot)$ is supported on the interval $(-1, 1)$ and is a symmetric probability density function with a bounded derivative. Furthermore, the Fourier transformation of $\widetilde{K}(\cdot)$ is absolutely integrable.

Assumption 2.1 is quite common in such problems. See Doukhan (1995) for the advantages of the geometric mixing. However, it would be possible, but with more tedious proofs, to obtain Theorems 2.1–2.3 under less restrictive assumptions that include some algebraically decaying rates. If $\{e_t\}$ is i.i.d. and $\{e_t\}$ is independent of $\{X_t\}$, then Assumption 2.1(i) requires only that the process $\{X_t\}$ is strictly stationary and α-mixing, and Assumption 2.1(ii) yields $E[e_t] = 0$ and $E[e_t^4] < \infty$. This is a natural condition in nonparametric autoregression. See, for example, Assumption 2.1 of Gao (1998). For the heteroscedastic case, we need to modify both Assumptions 2.1(i) and 2.1(ii). See, for example, Conditions (A2) and (A4) of Hjellvik, Yao and Tjøstheim (1998).

Assumption 2.2(i) is imposed to exclude the case that each g_i is already a linear combination of $\{f_{ij} : 1 \leq j \leq k_i\}$. For the case, model (2.2) is an additive polynomial regression. The choice of k_i is a model selection problem, which has already been discussed by Li (1987). The purpose of introducing Assumptions 2.2(i) and 2.2(ii) is to replace the unknown functions by finite series sums together with vectors of unknown parameters. Equation (2.57) is a standard smoothness condition in approximation theory. See Corollary 6.21 of Schumaker (1981) for the B-spline approximation, Chapter IV of Kashin and Saakyan (1989) for the trigonometric approximation, Theorem 0 of Gallant and Souza (1991) for the flexible Fourier form, and Chapter 7 of DeVore and Lorentz (1993) for the general orthogonal series approximation. If Assumption 2.2(ii) holds, then (2.58) is equivalent to

$$h_i^{2(m_i+1)} \int [F_i(u_i)^\tau \theta_i - g_i(u_i)]^2 p_i(u_i) du_i \approx C_i, \qquad (2.59)$$

where $\{p_i(u_i)\}$ denotes the marginal density function of V_{ti}. Equation

(2.59) is a standard smoothness condition in approximation theory. See Theorems 3.1 and 4.1 of Agarwal and Studden (1980) for the B–spline approximation and §3.2 of Eastwood and Gallant (1991) for the trigonometric approximation. This chapter extends existing results to the case where $\{V_{ti}\}$ is a strictly stationary process.

Assumption 2.3 is a kind of orthogonality condition, which holds when the process $\{X_t\}$ is strictly stationary and the series $\{f_{ij} : 1 \leq j \leq k_i, 1 \leq i \leq p\}$ are either in the family of trigonometric series or of Gallant's (1981) flexible Fourier form. For example, the orthogonality condition holds when $\{V_{t1}\}$ is strictly stationary and distributed uniformly over $[-1, 1]$ and $f_{1j}(V_{t1}) = \sin(j\pi V_{t1})$ or $\cos(j\pi V_{t1})$. Moreover, the orthogonality condition is a natural condition in nonparametric series regression. Assumption 2.4 is required to deal with this kind of problem. Many authors have used similar conditions. See, for example, (C.7) of Härdle and Vieu (1992).

Assumptions 2.5–2.8 are relatively mild in this kind of problem and can be justified in detail. Assumption 2.6(i) is necessary for the establishment of asymptotic normality in the semiparametric setting. As can be seen from Theorem 2.4, the condition on the existence of the inverse matrix, $\left(B^{UU}\right)^{-1}$, is required in the formulation of that theorem. Moreover, Assumption 2.6(i) corresponds to those used for the independent case. Assumption 2.6(ii) is needed as the existence of moments of higher than second order is required for this kind of problem when uniform convergence for nonparametric regression estimation is involved. Assumption 2.7(ii) is required due to the use of such a weight function. The continuity condition on the kernel function imposed in Assumption 2.8 is quite natural and easily satisfied.

Assumption 2.9 requires that when one of the bandwidths is proportional to $T^{-\frac{1}{5}}$, the optimal choice under a conventional criterion, the other bandwidths need to converge to zero with a rate related to $T^{-\frac{1}{5}}$. Assumption 2.9 is quite complex in general. However, it holds in some cases. For example, when we choose $p = 2$, $r = 3$, $\lambda = 4$, $a = 31$, $k = 1$, $b_1 = T^{-\frac{1}{5}}$, and $b_2 = T^{-\frac{2}{5}+\eta}$ for some $0 < \eta < \frac{1}{5}$, both (i) and (ii) hold. For instance,

$$\liminf_{T\to\infty} Tb_\pi^{\frac{2(r-1)a+2(\lambda r-2)}{(a+2)\lambda}} = \liminf_{T\to\infty} T^{\frac{19}{55}+\frac{12}{11}\eta} = \infty > 0.$$

and

$$\lim_{T\to\infty} Tb_\pi^{1+\frac{2}{r}} = \lim_{T\to\infty} T^{\frac{5}{3}\eta} = \infty.$$

Similarly to the independent case (Fan, Härdle and Mammen 1998, Remark 10), we assume that all the nonparametric components are only two times continuously differentiable and thus the optimal bandwidth b_k is proportional to $T^{-\frac{1}{5}}$. As a result, Assumption 2.9 basically implies $p \leq 4$. For example, in some cases the assumption of $p \leq 4$ is just sufficient for us to use an additive model to approximate the conditional mean $E[Y_t|Y_{t-1}, Y_{t-2}, Y_{t-3}, Y_{t-4}]$ by $g_1(Y_{t-1}) + g_2(Y_{t-2}) + g_3(Y_{t-3}) + g_4(Y_{t-4})$ with each $g_i(\cdot)$ being an unknown function. Nevertheless, we may ensure that the marginal integration method still works for the case of $p \geq 5$ and achieves the optimal rate of convergence by using a high–order kernel of the form

$$\int K(x)dx = 1, \quad \int x^i K(x)dx = 0 \text{ for } 1 \leq i \leq I - 1 \text{ and } \int x^I K(x) \neq 0$$
(2.60)

for $I \geq 2$ as discussed in Hengartner and Sperlich (2003) for the independent case, where I is the order of smoothness of the nonparametric components. This shows that in order to achieve the rate–optimal property, we will need to allow that smoothness increases with dimensions. This is well-known and has been used in some recent papers for the independent case (see Conditions A5, A7 and NW2–NW3 of Hengartner and Sperlich 2003).To ensure that the conclusions of the main results hold for this case, we need to slightly modify Assumptions 2.7–2.9. The details are similar to Assumptions 3.4'–3.6' of Gao, Lu and Tjøstheim (2006).

Assumption 2.10 is quite common in this kind of problem. Its justification may be available from the appendix of Xia, Tong and Li (1999).

2.5.2 Proofs of Theorems

This section provides the proofs of Theorems 2.3 and 2.4. The detailed proofs of the other theorems are referred to the relevant papers.

Proofs of Theorems 2.1 and 2.2: The detailed proofs are similar to those of Theorems 2.1 and 2.2 of Gao, Tong and Wolff (2002a).

Proof of Corollary 2.1: The proof follows from that of Theorem 2.2.

Proof of Theorem 2.3: We note that

$$\begin{aligned}
\widehat{\beta} - \beta &= \left(\frac{1}{T}\sum_{t=1}^{T}\widehat{U}_t^*\left(\widehat{U}_t^*\right)^{\tau}\right)^{-1}\left(\frac{1}{T}\sum_{t=1}^{T}\widehat{U}_t^*(\widehat{Y}_t^* - \widehat{U}_t^*\beta)\right) \\
&\equiv \left(B_T^{UU}\right)^{-1}B_T^{UY}.
\end{aligned}$$

Denote by $H_a^{(s)}(x) \equiv \sum_{l=1}^{p}P_{l,w}^{(s)}(x_l)$ and $H_a(x) \equiv \sum_{l=1}^{p}P_{l,w}^{U}(x_l)$ the additive approximate versions to $H^{(s)}(x) = E\left[\left(U_t^{(s)} - \mu_U^{(s)}\right)|V_t = x\right]$ and

$H(x) = E\left[(U_t - \mu_U)\,|V_t = x\right]$, respectively. We then define $H_{a,T}^{(s)}(x) \equiv \sum_{l=1}^{p} \widehat{P}_{l,w}^{(s)}(x_k)$ and $H_{a,T}(x) \equiv \sum_{l=1}^{p} \widehat{P}_{l,w}^{U}(x_l)$ as the corresponding estimators of $H_a^{(s)}(x)$ and $H_a(x)$. Then, we have

$$
\begin{aligned}
B_T^{UU} &= \frac{1}{T}\sum_{t=1}^{T}\left(\tilde{U}_t - H_a(V_t) + H_a(V_t) - H_{a,T}(V_t)\right) \\
&\quad \times \left(\tilde{Z}_t - H_a(V_t) + H_a(V_t) - H_{a,T}(V_t)\right)^{\tau} \\
&= \frac{1}{T}\sum_{t=1}^{T}\tilde{U}_t^*\left(\tilde{U}_t^*\right)^{\tau} + \frac{1}{T}\sum_{t=1}^{T}\tilde{U}_t^*\left(\Delta_t^{H_a}\right)^{\tau} \\
&\quad + \frac{1}{T}\sum_{t=1}^{T}\Delta_t^{H_a}\left(\tilde{U}_t^*\right)^{\tau} + \frac{1}{T}\sum_{t=1}^{T}\Delta_t^{H_a}\Delta_t^{H_a\,\tau} \equiv \sum_{l=1}^{4}B_{T,l}^{UU},
\end{aligned}
$$

where $\tilde{U}_t^* = \tilde{U}_t - H_a(V_t)$ and $\Delta_t^{H_a} = H_a(V_t) - H_{a,T}(V_t)$. Moreover,

$$
\begin{aligned}
B_T^{UY} &= \frac{1}{T}\sum_{t=1}^{T}\left(\tilde{U}_t - H_a(V_t) + H_a(V_t) - H_{a,T}(V_t)\right) \\
&\quad \times \left(\tilde{Y}_t - H_a^{(0)}(V_t) + H_a^{(0)}(V_t) - H_{a,T}^{(0)}(V_t)\right) \\
&\quad - \frac{1}{T}\sum_{t=1}^{T}\left(\tilde{U}_t - H_a(V_t) + H_a(V_t) - H_{a,T}(V_t)\right) \\
&\quad \times \left[\tilde{U}_t - H_a(V_t) + H_a(V_t) - H_{a,T}(V_t)\right]^{\tau}\beta \\
&= \frac{1}{T}\sum_{t=1}^{T}U_t^*\epsilon_t^* + \frac{1}{T}\sum_{t=1}^{T}U_t^*(\Delta_t^{(0)} - \Delta_t^{H_a\,\tau}\beta) \\
&\quad + \frac{1}{T}\sum_{t=1}^{T}\Delta_t^{H_a}\epsilon_t^* + \frac{1}{T}\sum_{t=1}^{T}\Delta_t^{H_a}\left[\Delta_t^{(0)} - \left(\Delta_t^{H_a}\right)^{\tau}\beta\right] \\
&\equiv \sum_{j=1}^{4}B_{T,j}^{UY}, \quad\quad\quad\quad\quad (2.61)
\end{aligned}
$$

where $\epsilon_t^* = Y_t^* - U_t^{*\tau}\beta$, U_t^* and $Y_t^* = \tilde{Y}_t - H_a^{(0)}(V_t)$ are as defined before, and $\Delta_t^{(s)} \equiv H_a^{(s)}(V_t) - H_{a,T}^{(s)}(V_t)$. So, to prove the asymptotic normality of $\widehat{\beta}$, it suffices to show that

$$
B_T^{UU} \xrightarrow{P} B^{UU} \quad \text{and} \quad \sqrt{T}\left(B_T^{UY} - \mu_B\right) \to_D N(0, \Sigma_B), \quad\quad (2.62)
$$

where B^{UU}, μ_B and Σ_B are as defined in Theorem 2.3. To this end, we need to have

$$\sum_{t=1}^{T} \left(\hat{P}_{k,w}^{(s)}(V_{tk}) - P_{k,w}^{(s)}(V_{tk}) \right)^2 = o_P(\sqrt{T}), \quad 0 \le s \le q, \qquad (2.63)$$

which follows from

$$\sup_{x_k \in [-L_k, L_k]} \left| \hat{P}_{k,w}^{(s)}(x_k) - P_{k,w}^{(s)}(x_k) \right| = O_P \left((Tb_k^{1+2/r})^{-r/(1+2r)} + b_k^2 \right)$$

$$+ \quad O_P(1) \sum_{l=1,\neq k}^{p} b_l^2 + O_P(1) \sum_{l=1,\neq k}^{p} b_l b_k$$

$$+ \quad o_P(1)b_k^2 + O_P(1) \left(\frac{1}{\sqrt{T}} \right) \qquad (2.64)$$

for some integer $r \ge 3$ and $Tb_k^5 = O(1)$,

$$\sqrt{T} \left((Tb_k^{1+2/r})^{-r/(1+2r)} + b_k^2 \right)^2 \le C \left(T(Tb_k^{1+2/r})^{-4r/(1+2r)} + Tb_k^8 \right)^{1/2}$$

$$= C \left(T^{-\frac{2r-1}{1+2r}} b_k^{-\frac{4(2+r)}{1+2r}} + Tb_k^8 \right)^{1/2} \to 0,$$

$$\sqrt{T} \left(\sum_{l=1,\neq k}^{p} b_l^2 + \sum_{l=1,\neq k}^{p} b_l b_k + b_k^2 + O_P(1) \left(\frac{1}{\sqrt{T}} \right) \right)^2 \to 0.$$

The proof of (2.63) is similar to that of Lemma 6.3 of Gao, Lu and Tjøstheim (2006).

Thus

$$\sum_{t=1}^{T} \left(\Delta_t^{(s)} \right)^2 = \sum_{t=1}^{T} \left(H_a^{(s)}(V_t) - H_{a,T}^{(s)}(V_t) \right)^2$$

$$= \sum_{t=1}^{T} \left(\sum_{k=1}^{p} (\hat{P}_{k,w}(V_{tk}) - P_{k,w}(V_{tk})) \right)^2$$

$$= o_P(\sqrt{T}). \qquad (2.65)$$

Therefore, using the Cauchy-Schwarz inequality, it follows that the (i,j)-th element of $B_{T,4}^{UU}$

$$B_{T,4}^{UU}(i,j) = \frac{1}{T} \sum_{t=1}^{T} \Delta_s^{(i)} \Delta_t^{(j)}$$

$$\le \frac{1}{T} \left(\sum_{t=1}^{T} (\Delta_s^{(i)})^2 \right)^{1/2} \left(\sum_{t=1}^{T} (\Delta_t^{(j)})^2 \right)^{1/2} = o_P(1),$$

and similarly

$$B_{T,2}^{UU}(i,j) = o_P(1) \text{ and } B_{T,3}^{UU}(i,j) = o_P(1).$$

Now, since $B_{T,1}^{UU} \to E\left[U_1^* U_1^{*\tau}\right]$ in probability, it follows from (2.61) that the first limit of (2.62) holds with $B^{UU} = E\left[U_1^* U_1^{*\tau}\right]$.

To prove the asymptotic normality in (2.62), by using the Cauchy–Schwarz inequality and (2.65), we have

$$\sqrt{T} \sum_{k=2}^{4} B_{T,k}^{UY} = o_P(1);$$

therefore, the second limit of (2.62) follows from (2.61) and

$$\sqrt{T}\left(B_{T,1}^{UY} - \mu_B\right) = \frac{1}{\sqrt{T}} \sum_{t=1}^{T} [U_t^* \epsilon_t^* - \mu_B] \to_D N(0, \Sigma_B)$$

with $\mu_B = E[R_t]$ and $\Sigma_B = E[R_0 R_t^\tau]$, where $R_t = U_t^* \epsilon_t^*$.

The proof of the asymptotic normality follows directly from the conventional central limit theorem for mixing time series. When (2.2) holds, the proof of the second half of Theorem 2.3 follows trivially.

Proof of Corollary 2.2: Its proof follows from that of Theorem 2.3.

Proof of Theorem 2.4: Note that

$$\widehat{\widehat{P}}_{k,w}(x_k) = \widehat{P}_{k,w}^{(0)}(x_k) - \widehat{\beta}^\tau \widehat{P}_{k,w}^{U}(x_k),$$

where $P_{k,w}(x_k) = P_{k,w}^{(0)}(x_k) - \beta^\tau P_{k,w}^{U}(x_k)$. Then

$$
\begin{aligned}
\widehat{\widehat{P}}_{k,w}(x_k) - P_{k,w}(x_k) &= [\widehat{P}_{k,w}^{(0)}(x_k) - P_{k,w}^{(0)}(x_k) \\
&\quad - \beta^\tau (\widehat{P}_{k,w}^{U}(x_k) - P_{k,w}^{U}(x_k))] \\
&\quad - (\widehat{\beta} - \beta)^\tau \widehat{P}_{k,w}^{U}(x_k) \\
&= P_{T,1}(x_k) + P_{T,2}(x_k).
\end{aligned}
$$

For any $c = (c_0, \ C_1^\tau)^\tau \in \mathbf{R}^{1+q}$ with $C_1 = (c_1, \cdots, c_q)^\tau \in \mathbf{R}^q$, we note that for $x_k \in [-L_k, L_k]$

$$
\begin{aligned}
\sum_{s=0}^{q} c_s P_{k,w}^{(s)}(x_k) &= c_0 P_{k,w}^{(0)}(x_k) + C_1^\tau P_{k,w}^{Z}(x_k) \\
&= c_0 E\left[H^{(0)}(V_t^{(-k)}, \ x_k)\right] w_{(-k)}(V_t^{(-k)}) \\
&\quad + C_1^\tau E\left[H^{(0)}(V_t^{(-k)}, \ x_k)\right] w_{(-k)}(V_t^{(-k)})
\end{aligned}
$$

$$= E\left[c_0 H^{(0)}(V_t^{(-k)}, \ x_k) + C_1^\tau H(V_t^{(-k)}, \ x_k)\right]$$

$$\times \ w_{(-k)}(V_t^{(-k)}) = E\left[g^{**}(V_t^{(-k)}, \ x_k)\right] \ w_{(-k)}(V_t^{(-k)}),$$

where $g^{**}(x) = E\left[Y_t^{**}|V_t = x\right]$ with $Y_t^{**} = c_0 \ (Y_t - \mu_Y) + C_1^\tau(U_t - \mu_U)$, and similarly

$$\sum_{s=0}^q c_s \widehat{P}_{k,w}^{(s)}(x_k) = c_0 \widehat{P}_{k,w}^{(0)}(x_k) + C_1^\tau \widehat{P}_{k,w}^U(x_k)$$

$$= \frac{1}{T} \sum_{t=1}^T g_T^{**}(V_t^{(-k)}, \ x_k) \ w_{(-k)}(V_t^{(-k)}),$$

where $g_T^{**}(x)$ is the local linear estimator of $g^{**}(x)$, as defined before with $\widetilde{Y}_t^{**} = c_0\widetilde{Y}_t + C_1^\tau\widetilde{U}_t$ instead of $\widetilde{Y}_t$ there. Therefore, similarly to the proof of Lemma 6.5 of Gao, Lu and Tjøstheim (2006), the distribution of

$$\sqrt{Tb_k} \sum_{s=0}^q c_s \left(\widehat{P}_{k,w}^{(s)}(x_k) - P_{k,w}^{(s)}(x_k)\right) \tag{2.66}$$

is asymptotically normal.

Now taking $c_0 = 0$ in (2.66) shows that $\widehat{P}_{k,w}^U(x_k) \rightarrow P_{k,w}^U(x_k)$ in probability, which together with Theorem 2.3 leads to

$$\sqrt{T \, b_k} \, P_{T,2}(x_k) = \sqrt{Tb_k} \, (\widehat{\beta} - \beta)^\tau \widehat{P}_{k,w}^U(x_k)$$

$$= O_P(\sqrt{b_k}) = o_P(1). \tag{2.67}$$

On the other hand, taking $c_0 = 1$ and $C_1 = -\beta$ in (2.66), we have

$$\sqrt{Tb_k} \, P_{T,1}(x_k) = \sqrt{Tb_k} \left[\widehat{P}_{k,w}^{(0)}(x_k) - P_{k,w}^{(0)}(x_k)\right]$$

$$- \sqrt{Tb_k} \, \beta^\tau \left(\widehat{P}_{k,w}^U(x_k) - P_{k,w}^U(x_k)\right) \tag{2.68}$$

are asymptotically normally distributed with $Y_t^{**} = Y_t - \mu_Y - \beta^\tau(U_t - \mu_U)$ and $g^{**}(x) = E(Y_t^{**}|V_t = x)$. This finally yields Theorem 2.4.

Proofs of Theorems 2.5 and 2.6: The detailed proofs are the same as those of Theorems 2 and 4 of Xia, Tong and Li (1999), respectively.

2.6 Bibliographical notes

Further to the nonparametric kernel estimation method discussed in this chapter, some related methods have been considered in Linton and Härdle (1996), Linton (1997. 2000, 2001), Li and Wooldridge (2002), Xia *et al.* (2002), Yang (2002), Yang and Tschernig (2002), Huang and Yang (2004), Horowitz and Mammen (2004), Linton and Mammen (2005), Yang (2006), and others.

Using orthogonal series and spline smoothing methods in econometrics and statistics has a long history. Eumunds and Moscatelli (1977), Wahba (1978), Agarwal and Studden (1980), and Gallant (1981) were among the first to use spline smoothing and trigonometric approximation methods in nonparametric regression. Since their studies for the fixed designs and independent errors, several authors have extended such estimation methods to nonparametric and semiparametric regression models with both random designs and time series errors. Recent extensions include Eubank (1988), Härdle (1990), Andrews (1991), Gao and Liang (1995), Fan and Gijbels (1996), Gao and Liang (1997), Gao and Shi (1997), Gao (1998), Eubank (1999), Shi and Tsai (1999), Gao, Tong and Wolff (2002a, 2002b), Fan and Li (2003), Hyndman *et al.* (2005), and others.

In addition to nonparametric kernel, orthogonal series and wavelet methods, some other nonparametric methods, such as the empirical likelihood method and profile likelihood method proposed for estimating nonparametric and semiparametric regression models with independent designs and errors, could also be applicable to nonparametric and semiparametric time series regression. Recent studies include Cai, Fan and Li (2000), Cai, Fan and Yao (2000), Chen, Härdle and Li (2003), Fan and Huang (2005), and Fan and Jiang (2005).

This chapter has concentrated on the case where both $\{X_t\}$ and $\{e_t\}$ are stationary time series. Recent studies show that it is possible to apply the kernel method to deal with the case where $\{X_t\}$ may not be stationary. To the best of our knowledge, Granger, Inoue and Morin (1997) , Karlsen and Tjøstheim (1998), and Phillips and Park (1998) were among the first to apply the kernel method to estimate the conditional mean function of Y_t given $X_t = x$ when $\{X_t\}$ is nonstationary. Further studies have been given in Karlsen and Tjøstheim (2001) and Karlsen, Myklebust and Tjøstheim (2006).

CHAPTER 3

Nonlinear Time Series Specification

3.1 Introduction

Let (Y, X) be a $d + 1$–dimensional vector of random variables with Y the response variable and X the vector of d–dimensional covariates. We assume that both X and Y are continuous random variables with $\pi(x)$ as the marginal density function of X, $f(y|x)$ being the conditional density function of Y given $X = x$ and $f(x, y)$ as the joint density function. Let $\mu_j(x) = E[Y^j | X = x]$ denote the j–th conditional moment of Y given $X = x$. Let $\{(Y_t, X_t) : 1 \leq t \leq T\}$ be a sequence of observations drawn from the joint density function $f(x, y)$. As the three density functions may not be known parametrically, various nonparametric estimation methods have been proposed in the literature (see Silverman 1986; Wand and Jones 1995; Fan and Gijbels 1996; Fan and Yao 2003; and others).

In recent years, nonparametric and semiparametric techniques have been used to construct model specification tests for $\mu_j(x)$. Interest focuses on tests for a parametric form versus a nonparametric form, tests for a semiparametric (partially linear or single–index) form against a nonparametric form, and tests for the significance of a subset of the nonparametric regressors. Härdle and Mammen (1993) were among the first to develop consistent tests for parametric specification by employing the kernel regression estimation technique. There have since been many advances. By contrast, there are only several papers available to the best of our knowledge in the field of parametric specification of density functions. Aït-Sahalia (1996b) was among the first to propose specifying marginal density functions parametrically. Pritsker (1998) evaluated the performance of Aït–Sahalia's test and concluded that the usage of asymptotic normality as the first–order approximation to the distribution of the proposed test may significantly contribute to the poor performance of the proposed test in the finite–sample analysis. Gao and King (2004) proposed a much improved test than the Aït–Sahalia's test for parametric specification of marginal density functions. More recently, Chen and Gao

(2005) developed an empirical–likelihood driven test statistic for parametric specification of the conditional density function of $f(y|x)$ with application in diffusion process specification. Hong and Li (2005) proposed using two nonparametric transition density–based tests for continuous–time diffusion models.

Apart from using such test statistics based on nonparametric kernel, nonparametric series, spline smoothing and wavelet methods, there are various test statistics constructed and studied based on empirical distributions. Such studies include Andrews (1997), Stute (1997), Stute, Thies and Zhu (1998), Whang (2000), Stute and Zhu (2002, 2005), Zhu (2005), and others. A different class of test statistics based on the so–called pseudo–likelihood ratio and generalized likelihood ratio have also been studied extensively mainly for cases of fixed designs and independent errors. Recent studies in this field include Fan and Huang (2001), Fan and Yao (2003), Fan and Zhang (2003), Chen and Fan (2005), Fan and Huang (2005), Fan and Jiang (2005), and others.

Since the literature on nonparametric and semiparametric specification testing is huge, we concentrate on parametric specification testing of the conditional mean function $\mu_1(x) = E[Y|X = x]$ and the conditional variance function $\sigma^2(x) = \mu_2(x) - \mu_1^2(x)$ in this chapter. The rest of this chapter is organised as follows. Section 3.2 discusses existing tests for conditional mean functions and then demonstrates that various existing nonparametric kernel tests can be decomposed with each of the leading terms being a quadratic form of dependent time series. Section 3.3 briefly considers semiparametric testing for conditional variance functions. Some general semiparametric testing problems are also discussed in Section 3.4. Section 3.5 presents an example of implementation. Mathematical assumptions and proofs are relegated to Section 3.6.

3.2 Testing for parametric mean models

Consider a nonlinear time series model of the form

$$Y_t = m(X_t) + e_t, \ t = 1, 2, \ldots, T, \tag{3.1}$$

where $\{X_t\}$ is a sequence of strictly stationary time series variables, $\{e_t\}$ is a sequence of independent and identically distributed (i.i.d.) errors with $E[e_t] = 0$ and $0 < E[e_t^2] = \sigma^2 < \infty$, $m(\cdot)$ is an unknown function defined over $\mathbf{R}^d = (-\infty, \infty)^d$ for $d \geq 1$, and T is the number of observations. Moreover, we assume that $\{X_s\}$ and $\{e_t\}$ are independent for all $1 \leq s \leq t \leq T$ and that the distribution of $\{e_t\}$ may be unknown nonparametrically or semiparametrically.

To avoid the so–called curse of dimensionality problem, this chapter mainly considers the case of $1 \leq d \leq 3$. For higher dimensional cases, Section 3.4 discusses several dimension reduction procedures.

In recent years, nonparametric and semiparametric techniques have been used to construct model specification tests for the mean function of model (3.1). Interest focuses on tests for a parametric form versus a nonparametric form, tests for a semiparametric (partially linear or single–index) form against a nonparametric form, and tests for the significance of a subset of the nonparametric regressors.

Among existing nonparametric and semiparametric tests, an estimation based optimal choice of either a bandwidth value, such as cross–validation selected bandwidth, or a truncation parameter based on a generalized cross–validation selection method is used in the implementation of each of the proposed tests. Nonparametric tests involving the second approach of choice of either a set of suitable bandwidth values for the kernel case or a sequence of positive integers for the smoothing spline case include Fan (1996), Fan, Zhang and Zhang (2001), Horowitz and Spokoiny (2001), Chen and Gao (2004, 2005), and Arapis and Gao (2006). The practical implementation of choosing such sets or sequences is, however, problematic. This is probably why Horowitz and Spokoiny (2001) developed their theoretical results based on a set of suitable bandwidths on the one hand but choose their practical bandwidth values based on the assessment of the power function of their test on the other hand.

Recently, some new approaches have been discussed to support such a power-based bandwidth selection procedure. On the issue of size correction, to the best of our knowledge, the only available paper on parametric specification of model (3.1) is given by Fan and Linton (2003), who developed an Edgeworth expansion for the size function of their test. Some other related studies include Nishiyama and Robinson (2000), Horowitz (2003), and Nishiyama and Robinson (2005), who established some useful Edgeworth expansions for bootstrap distributions of partial–sum type of tests for improving the size performance.

More recently, Gao and Gijbels (2006) have discussed a sound approach to choosing smoothing parameters in nonparametric and semiparametric testing. The main idea is to find an Edgeworth expansion of the asymptotic distribution of the test concerned. Due to the involvement of such smoothing parameters in the Edgeworth expansion, the authors have been able to explicitly express the leading terms of both the size and power functions and then determine how the smoothing parameters should be chosen according to certain requirements for both the size and power functions. For example, when a significance level is given, the authors have chosen the smoothing parameters such that the size function is controlled by the significance level while the power function is maximized. Both novel theory and methodology are established. In addition, the authors have then developed an easy implementation procedure for the practical realization of the established methodology. In the rest of this section, we discuss the main results given in Gao and Gijbels (2006).

The main interest of this section is to test

$$
\begin{aligned}
\mathcal{H}_{01} : m(x) &= m_{\theta_0}(x) \quad \text{versus} \\
\mathcal{H}_{11} : m(x) &= m_{\theta_1}(x) + C_T \Delta(x) \quad \text{for all } x \in \mathbf{R}^d, \quad (3.2)
\end{aligned}
$$

where $\theta_0, \theta_1 \in \Theta$ with Θ being a parameter space of $\mathbf{R}^d$, C_T is a sequence of real numbers and $\Delta(x)$ is a continuous function over $\mathbf{R}^d$.

Under $\mathcal{H}_{01}$, model (3.1) becomes a semiparametric time series model of the form

$$
Y_t = m_{\theta_0}(X_t) + e_t \quad (3.3)
$$

when the distribution of $\{e_t\}$ is unknown nonparametrically or semiparametrically.

3.2.1 Existing test statistics

Härdle and Mammen (1993) introduced the L_2–distance between a nonparametric kernel estimator of $m(\cdot)$ and a parametric counterpart. More precisely, let us denote the nonparametric estimator of $m(\cdot)$ by $\widehat{m}_h(\cdot)$ and the parametric estimator of $m_\theta(\cdot)$ by $\widetilde{m}_{\widehat{\theta}}(\cdot)$ and consider

$$
\begin{aligned}
\widehat{m}_h(x) &= \frac{\sum_{t=1}^T K_h(x - X_t) Y_t}{\sum_{t=1}^T K_h(x - X_t)} \\
\widetilde{m}_{\widehat{\theta}}(x) &= \frac{\sum_{t=1}^T K_h(x - X_t) m_{\widehat{\theta}}(X_t)}{\sum_{t=1}^T K_h(x - X_t)}, \quad (3.4)
\end{aligned}
$$

where $\widehat{\theta}$ is a $\sqrt{T}$–consistent estimator of θ_0 under $\mathcal{H}_{01}$, $K_h(\cdot) = \frac{1}{h^d} K\left(\frac{\cdot}{h}\right)$, with $K(\cdot)$ being the probability kernel density function and h being the bandwidth parameter.

Härdle and Mammen (1993) proposed using a test statistic of the form

$$
\begin{aligned}
M_{T1}(h) &= T h^{\frac{d}{2}} \int \left\{ \widehat{m}_h(x) - \widetilde{m}_{\widehat{\theta}}(x) \right\}^2 w(x) dx \\
&= T h^{\frac{d}{2}} \int \left(\frac{\left[\sum_{t=1}^T K_h(x - X_t) \left(Y_t - m_{\widehat{\theta}}(X_t) \right) \right]^2}{T^2 \widehat{\pi}^2(x)} \right) w(x) dx
\end{aligned}
$$

$$
\begin{aligned}
=\quad & Th^{\frac{d}{2}}\sum_{t=1}^{T}\sum_{s=1}^{T}\left(\int\frac{K_h(x-X_s)K_h(x-X_t)}{T^2\widehat{\pi}^2(x)}w(x)dx\right)e_s e_t \\
-\quad & 2Th^{\frac{d}{2}}\sum_{s=1}^{T}\sum_{t=1}^{T}\left(\int\frac{K_h(x-X_s)K_h(x-X_t)}{T^2\widehat{\pi}^2(x)}w(x)dx\right)e_s \\
\times\quad & \left[m_{\widehat{\theta}}(X_t)-m(X_t)\right] \\
+\quad & Th^{\frac{d}{2}}\sum_{s=1}^{T}\sum_{t=1}^{T}\left(\int\frac{K_h(x-X_s)K_h(x-X_t)}{T^2\widehat{\pi}^2(x)}w(x)dx\right) \\
\times\quad & \left[m_{\widehat{\theta}}(X_s)-m(X_t)\right]\cdot\left[m_{\widehat{\theta}}(X_t)-m(X_t)\right],
\end{aligned}
\tag{3.5}
$$

where $\widehat{\pi}(x)=\frac{1}{T}\sum_{t=1}^{T}K_h(x-X_t)$ is the kernel density estimator of the marginal density function, $\pi(x)$, of $\{X_t\}$, and $w(\cdot)$ is some nonnegative weight function. The first term on the right–hand side of (3.5) is the leading term of $M_{T1}(h)$. Moreover, under $\mathcal{H}_{01}$ the asymptotic normality of $M_{T1}(h)$ follows from a central limit theorem for such a quadratic representation of the first term.

Since $\widehat{\pi}(x)$ is an asymptotically consistent estimator of $\pi(x)$, we may replace $\widehat{\pi}(x)$ by $\pi(x)$ in $M_{T1}(h)$. This implies a test statistic of the form

$$
\begin{aligned}
M_{T2}(h)\quad=\quad & Th^{\frac{d}{2}}\int\left(\sum_{t=1}^{T}K_h(x-X_t)\left[Y_t-m_{\widehat{\theta}}(X_t)\right]\right)^2 w(x)dx \\
=\quad & Th^{\frac{d}{2}}\sum_{t=1}^{T}\sum_{s=1}^{T}\left(\int K_h(x-X_s)K_h(x-X_t)w(x)dx\right)e_s e_t \\
-\quad & 2Th^{\frac{d}{2}}\sum_{s=1}^{T}\sum_{t=1}^{T}\left(\int K_h(x-X_s)K_h(x-X_t)w(x)dx\right)e_s \\
\times\quad & \left[m_{\widehat{\theta}}(X_t)-m(X_t)\right] \\
+\quad & Th^{\frac{d}{2}}\sum_{t=1}^{T}\sum_{t=1}^{T}\left(\int K_h(x-X_s)K_h(x-X_t)w(x)dx\right) \\
\times\quad & \left[m_{\widehat{\theta}}(X_s)-m(X_s)\right]\cdot\left[m_{\widehat{\theta}}(X_t)-m(X_t)\right],
\end{aligned}
\tag{3.6}
$$

as studied in Kreiss, Neumann and Yao (2002), where $w(\cdot)$ is a nonnegative weight function probably depending on $\pi(\cdot)$.

Kreiss, Neumann and Yao (2002) established the asymptotic normality of $M_{T2}(h)$ based on the leading quadratic term in (3.6). In the implementation, they developed a wild–bootstrap procedure for finding an approximate critical value. The choice of h is also based on an optimal estimation selection criterion.

As an alternative to $M_{T1}(h)$, Horowitz and Spokoiny (2001) used a discrete approximation to $M_{T1}(h)$ of the form

$$M_{T3}(h) = \sum_{t=1}^{T} \left(\widehat{m}_h(X_t) - \widetilde{m}_{\widehat{\theta}}(X_t) \right)^2, \tag{3.7}$$

where $\{X_t\}$ is only a sequence of fixed designs. They further considered a multiscale normalized version of the form

$$M_{T3} = \max_{h \in H_{3T}} \frac{M_{T3}(h) - \widehat{M}_T(h)}{\widehat{V}_T(h)}, \tag{3.8}$$

where H_{3T} is a set of suitable bandwidths,

$$\widehat{M}_T(h) = \sum_{t=1}^{T} \left(\sum_{s=1}^{T} W_h(X_s, X_t) \right) \widehat{\sigma}_T^2(X_t),$$

and

$$\widehat{V}_T^2(h) = 2 \sum_{s=1}^{T} \sum_{t=1}^{T} \left(\sum_{\ell=1}^{T} W_h(X_\ell, X_s) W_h(X_\ell, X_t) \right)^2 \widehat{\sigma}_T^2(X_s) \widehat{\sigma}_T^2(X_t),$$

in which $W_h(\cdot, X_t) = \frac{K_h(\cdot - X_t)}{\sum_{u=1}^{T} K_h(\cdot - X_u)}$ and $\widehat{\sigma}_T^2(X_s)$ is a consistent estimator of the variance function $\sigma^2(X_t) = E[e_t^2]$. Horowitz and Spokoiny (2001) then showed that M_{T3} is asymptotically consistent with an optimal rate of convergence for testing. Theoretically, certain conditions are imposed on H_{3T} for the technical proofs on the one hand, but their practical bandwidth values are chosen based on the assessment of the power function of their test (in simulations) on the other hand. This was part of the motivation used in Gao and Gijbels (2006) to propose a radical approach to optimally choosing the suitable bandwidth based on the assessment of the power function of the test under consideration.

As can be seen from the construction of $M_{Ti}(h)$ for $i = 1, 2, 3$, a secondary estimation procedure is normally required for $\sigma^2(\cdot)$ and some other higher–order moments when the variance function is not constant. To avoid such a secondary estimation procedure, Chen, Härdle and Li (2003) proposed a test statistic based on empirical likelihood ideas. As shown in their paper, the first–order approximation of their test is asymptotically equivalent to

$$M_{T4}(h) = Th^d \int \frac{\left[\widehat{m}_h(x) - \widetilde{m}_{\widehat{\theta}}(x) \right]^2}{V(x)} dx$$

$$= Th^d \int \frac{\left[\sum_{t=1}^T K_h(x - X_t)\left(Y_t - m_{\widehat{\theta}}(X_t)\right)\right]^2}{T^2 \widehat{\pi}^2(x) V(x)} dx,$$

where $V(x) = \frac{\sigma^2(x)}{\pi(x)} \int K^2(u)du$. As shown in Fan and Gijbels (1996), $\frac{1}{Th^d}V(x)$ behaves as the asymptotic variance of the Nadaraya–Watson estimator.

Applying a recently developed central limit theorem for degenerate $U-$ statistics, Chen, Härdle and Li (2003) derived the asymptotic normality of their test. With respect to the choice of h in practice, they suggested choosing h based on any bandwidth selector which minimizes the mean squared error of the nonparametric curve estimation.

Recently, Chen and Gao (2004) combined the scheme of Horowitz and Spokoiny (2001) with the empirical likelihood feature of Chen, Härdle and Li (2003) to propose a novel kernel test statistic for testing (3.2). Their experience shows that the choice of H_{T3} in the finite–sample case can be quite arbitrary and even problematic.

So far we have briefly established the first group of nonparametric kernel test statistics based on the L_2-distance function between a nonparametric kernel estimator and a parametric counterpart of the mean function. In addition, we have also mentioned that under $\mathcal{H}_{01}$ the leading term of each of the tests $M_{Ti}(h)$ for $1 \leq i \leq 4$ is of a quadratic form

$$P_T(h) = \sum_{t=1}^T \sum_{s=1}^T e_s \, w(X_s) L_h(X_s - X_t) w(X_t) \, e_t, \qquad (3.9)$$

where $L_h(\cdot) = \frac{1}{T\sqrt{h^d}} L\left(\frac{\cdot}{h}\right)$, $L(x) = \int K(y)K(x + y)dy$, and $w(\cdot)$ is a suitable weight function probably depending on either $\pi(\cdot)$, $\sigma^2(\cdot)$ or both.

In the following, we construct the second group of nonparametric kernel test statistics using a different distance function. We now rewrite model (3.1) into a notational version of the form under $\mathcal{H}_{01}$

$$Y = m_{\theta_0}(X) + e, \qquad (3.10)$$

where X is assumed to be random and θ_0 is the true value of θ under $\mathcal{H}_{01}$. Obviously, $E[e|X] = 0$ under $\mathcal{H}_{01}$. Existing studies (Zheng 1996; Li and Wang 1998; Li 1999; Fan and Linton 2003) have proposed using a distance function of the form

$$E\left[eE\left(e|X\right)\pi(X)\right] = E\left[\left(E^2(e|X)\right)\pi(X)\right], \qquad (3.11)$$

where $\pi(\cdot)$ is the marginal density function of X.

This would suggest using a normalized kernel–based sample analogue of

(3.11) of the form

$$
\begin{aligned}
L_T(h) &= \frac{h^{\frac{d}{2}}}{T} \sum_{s=1}^{T}\sum_{t=1}^{T} \widehat{e}_s \, K_h(X_s - X_t)\, \widehat{e}_t \\
&= \frac{h^{\frac{d}{2}}}{T} \sum_{s=1}^{T}\sum_{t=1}^{T} e_s \, K_h(X_s - X_t)\, e_t \\
&+ \frac{2h^{\frac{d}{2}}}{T} \sum_{s=1}^{T}\sum_{t=1}^{T} e_s \, K_h(X_s - X_t)\left[m(X_t) - m_{\widehat{\theta}}(X_t)\right] \\
&+ \frac{h^{\frac{d}{2}}}{T} \sum_{s=1}^{T}\sum_{t=1}^{T} K_h(X_s - X_t)\left[m(X_s) - m_{\widehat{\theta}}(X_s)\right] \\
&\times \left[m(X_t) - m_{\widehat{\theta}}(X_t)\right],
\end{aligned}
\tag{3.12}
$$

where $\widehat{e}_t = Y_t - m_{\widehat{\theta}}(X_t)$. It can easily be seen that under $\mathcal{H}_{01}$ the leading term of $L_T(h)$ is of a quadratic form

$$
\begin{aligned}
Q_T(h) &= \frac{h^{\frac{d}{2}}}{T} \sum_{s=1}^{T}\sum_{t=1}^{T} e_s \, K_h(X_s - X_t)\, e_t \\
&= \frac{1}{T\sqrt{h^d}} \sum_{s=1}^{T}\sum_{t=1}^{T} e_s \, K\left(\frac{X_s - X_t}{h}\right) e_t.
\end{aligned}
\tag{3.13}
$$

Various versions of (3.12) have been used in Fan and Li (1996), Zheng (1996), Li and Wang (1998), Li (1999), Fan and Linton (2003), Gao and King (2005), and others. Recently, Zhang and Dette (2004) compared both large and finite–sample properties of $M_{T1}(h)$, $L_T(h)$ and the main test proposed in Fan, Zhang and Zhang (2001). Their large sample study shows that the main conditions required to impose on h are $\lim_{T\to\infty} h = 0$ and $\lim_{T\to\infty} Th^d = \infty$, which are the minimal conditions for establishing asymptotic normality. Their finite–sample study is then based on the choice $h \sim T^{-\frac{2}{4d+1}}$.

In summary, Equations (3.9) and (3.13) can be generally written as

$$
\begin{aligned}
R_T(h) &= \sum_{s=1}^{T}\sum_{t=1}^{T} e_s \, \phi_T(X_s, X_t)\, e_t \\
&= \sum_{s=1}^{T} \phi_T(X_s, X_s) e_s^2 + \sum_{s=1}^{T}\sum_{t=1,\neq t}^{T} \phi_T(X_s, X_t) e_s e_t,
\end{aligned}
\tag{3.14}
$$

where the quantity $\phi_T(\cdot,\cdot)$ always depends on the sample size T, the bandwidth h and the kernel function K.

As briefly discussed below, the leading term of each of some other existing nonparametric kernel tests can also be represented by a quadratic form of the type (3.14). Thus, it is of general interest to study asymptotic distributions and their Edgeworth expansions for such quadratic forms. To present the main idea of establishing Edgeworth expansions for such quadratic forms, we concentrate on $Q_T(h)$ and $L_T(h)$ in Sections 3.2.2 and 3.2.3 below. This is because the main technology for establishing an Edgeworth expansion for the asymptotic distribution of any of such tests is the same as that for $Q_T(h)$, although more technicalities may be needed for some individual cases.

3.2.2 Asymptotic distributions and expansions

The aim of this section is to get clear theoretical insights into the problem of smoothing parameter selection in the testing context. In this section, we establish some novel results in theory before we will discuss how to realize such theoretical results in practice.

Before we establish an Edgeworth expansion for the asymptotic distribution of $Q_n(h)$ defined in (3.13), we need to introduce the following notation: Let $\mu_k = E[e_1^k]$ for $1 \le k \le 6$ and $\nu_l = E[\pi^l(X_1)]$ $(1 \le l \le 3)$;

$$
\sigma_T^2 = (\mu_4 - \mu_2^2) \frac{K^2(0)}{Th^d} + 2\mu_2^2 \nu_2 \int K^2(u) du \text{ and}
$$

$$
\kappa_T = \frac{\sqrt{h^d} \left(\frac{\mu_3^2 K^2(0)}{Th^d} + \frac{4\mu_2^3 \nu_3}{3} K^{(3)}(0) \right)}{\sigma_T^3}, \tag{3.15}
$$

where $K^{(3)}(\cdot)$ is the three–time convolution of $K(\cdot)$ with itself, $\nu_l = E[\pi^l(X_1)] = \int \pi^{l+1}(x) dx$ where $\pi(\cdot)$ is the marginal density function.

The proof of Theorem 3.1 below is relegated to Section 3.5 of this chapter.

Theorem 3.1. *Suppose that Assumption 3.1 listed in Section 3.5 below holds. Then*

$$
\sup_{x \in \mathbf{R}^1} \left| P\left(\frac{Q_T(h) - E[Q_T(h)]}{\sigma_T} \le x \right) - \Phi(x) + \kappa_T \left(x^2 - 1 \right) \phi(x) \right| \le Ch^d,
$$

$$\tag{3.16}$$

where $\phi(x)$ and $\Phi(x)$ denote the respective probability density function and cumulative distribution function of the standard normal random variable, and $0 < C < \infty$ is an absolute constant.

It follows from Theorem 3.1 that as $h \to 0$ and $Th^d \to \infty$

$$\sup_{x \in \mathbf{R}^1} \left| P\left(\frac{Q_T(h) - E[Q_T(h)]}{\sigma_T} \leq x \right) - \Phi(x) + \kappa_T \left(x^2 - 1\right) \phi(x) \right| \to 0.$$

(3.17)

This shows that Theorem 3.1 is of importance and usefulness in itself, since existing central limit theorems established in both the econometrics and statistics literature have implied only the standard asymptotic normality for the normalized version of $Q_T(h)$ in (3.13).

To study both the size and power properties of $\widehat{L}_T(h) = \frac{L_T(h) - E[L_T(h)]}{\sqrt{\mathrm{var}[L_T(h)]}}$,

we first need to find an approximate α–level critical value for $\widehat{L}_T(h)$. For each given h, we define a stochastically normalized version of the form

$$\overline{L}_T(h) = \frac{\sum_{s=1}^{T} \sum_{t=1, \neq x}^{T} \widehat{e}_s \, K_h(X_s - X_t) \, \widehat{e}_t}{\sqrt{2 \sum_{s=1}^{T} \sum_{t=1}^{T} \widehat{e}_s^2 \, K_h^2(X_s - X_t) \, \widehat{e}_t^2}}.$$

(3.18)

As pointed out in existing literature (such as, Zheng 1996; Li and Wang 1998; Fan and Linton 2003; Casas and Gao 2005; Arapis and Gao 2006), the version $\overline{L}_T(h)$ has three main features: (i) it appears to be more straightforward computationally; (ii) it is invariant to $\sigma^2 = \mu_2^2 = E[e_t^2]$; and (iii) there is no need to estimate any moments higher than σ^2. When σ^2 is unknown, it is estimated by $\widehat{\sigma}_T^2 = \frac{1}{T} \sum_{t=1}^{T} \widehat{e}_t^2$.

In addition, it can be easily shown that $\overline{L}_T(h) = \widehat{L}_T(h) + o_P(1)$ for each given h. Thus, we may use the distribution of $\overline{L}_T(h)$ to approximate that of $\widehat{L}_T(h)$. Let l_α^e ($0 < \alpha < 1$) be the $1 - \alpha$ quantile of the exact finite–sample distribution of $\widehat{L}_T(h)$. Because l_α^e may not be evaluated in practice, we therefore choose either a nonrandom approximate α–level critical value, l_α, or a stochastic approximate α–level critical value, l_α^*, by using the following simulation procedure:

- Since $\overline{L}_T(h)$ is invariant to σ^2, we generate $Y_t^* = m_{\widehat{\theta}}(X_t) + e_t^*$ for $1 \leq t \leq T$, where $\{e_t^*\}$ is a sequence of independent and identically distributed random samples drawn from a prespecified distribution, such as $N(0,1)$. Use the data set $\{(X_t, Y_t^*) : 1 \leq t \leq T\}$ to estimate $\widehat{\theta}$ by $\widehat{\theta}^*$ and compute $\overline{L}_T(h)$. Let l_α be the $1 - \alpha$ quantile of the distribution of

$$\overline{L}_T^*(h) = \frac{\sum_{s=1}^{T} \sum_{t=1, \neq s}^{T} \widehat{e}_s^* \, K_h(X_s - X_t) \, \widehat{e}_t^*}{\sqrt{2 \sum_{s=1}^{T} \sum_{t=1}^{T} \widehat{e}_s^{*2} \, K_h^2(X_s - X_t) \, \widehat{e}_t^{*2}}},$$

where $\widehat{e}_s^* = Y_s^* - m_{\widehat{\theta}^*}(X_s)$. In the simulation process, the original sample $\mathcal{X}_T = (X_1, \cdots, X_T)$ acts in the resampling as a fixed design even when $\{X_t\}$ is a sequence of random variables.

- Repeat the above step M times and produce M versions of $\overline{L}_T^*(h)$ denoted by $\overline{L}_{T,m}^*(h)$ for $m = 1, 2, \ldots, M$. Use the M values of $\overline{L}_{T,m}^*(h)$ to construct their empirical distribution function. The bootstrap distribution of $\overline{L}_T^*(h)$ given $\mathcal{W}_T = \{(X_t, Y_t) : 1 \leq t \leq T\}$ is defined by $P^* \left(\overline{L}_T^*(h) \leq x \right) = P \left(\overline{L}_T^*(h) \leq x | \mathcal{W}_T \right)$. Let l_α^* $(0 < \alpha < 1)$ satisfy $P^* \left(\overline{L}_T^*(h) \geq l_\alpha^* \right) = \alpha$ and then estimate l_α by l_α^*.

Note that both $l_\alpha = l_\alpha(h)$ and $l_\alpha^* = l_\alpha^*(h)$ depend on h. Let $\widehat{L}_T^*(h)$ be the corresponding version of $\widehat{L}_T(h)$ when the bootstrap resamples are used. Note also that the above simulation is based on the so–called regression bootstrap simulation procedure discussed in the literature, such as Li and Wang (1998), Kreiss, Neumann and Yao (2002), and Franke, Kreiss and Mammen (2002). When $X_t = Y_{t-1}$, we may also use a recursive simulation procedure, which is another commonly-used simulation procedure in the literature. See, for example, Hjellvik and Tjøstheim (1995), Franke, Kreiss and Mammen (2002), and others. In addition, we may also use a wild bootstrap to generate a sequence of resamples for $\{e_t^*\}$.

Since the choice of a simulation procedure does not affect the establishment of the main results of this chapter, they are thus stated based on the proposed simulation procedure.

We now have the following results; their proofs are given in Section 3.5 of this chapter.

Theorem 3.2. (i) *Suppose that Assumptions 3.1 and 3.2 listed in Section 3.5 below hold. Then under $\mathcal{H}_{01}$*

$$\sup_{x \in \mathbf{R}^1} \left| P^*(\widehat{L}_T^*(h) \leq x) - P(\widehat{L}_T(h) \leq x) \right| = O\left(\sqrt{h^d} \right) \tag{3.19}$$

holds in probability with respect to the joint distribution of $\mathcal{W}_T$; and

$$P \left(\widehat{L}_T(h) > l_\alpha^* \right) = \alpha + O\left(\sqrt{h^d} \right). \tag{3.20}$$

(ii) *Suppose that Assumptions 3.1–3.3 listed in Section 3.5 below hold. Then under $\mathcal{H}_{11}$*

$$\lim_{T \to \infty} P \left(\widehat{L}_T(h) > l_\alpha^* \right) = 1. \tag{3.21}$$

For a similar test statistic, Li and Wang (1998) established some results weaker than (3.19). Fan and Linton (2003) considered some higher–order approximations to the size function of the test discussed in Li and Wang (1998).

3.2.3 Size and power functions

For each h we define the following size and power functions

$$\alpha_T(h) = P\left(\widehat{L}_T(h) > l_\alpha | \mathcal{H}_{01} \text{ holds}\right) \text{ and}$$

$$\beta_T(h) = P\left(\widehat{L}_T(h) > l_\alpha | \mathcal{H}_{11} \text{ holds}\right). \qquad (3.22)$$

Correspondingly, we define $(\alpha_T^*(h), \beta_T^*(h))$ with l_α replaced by l_α^*.

As discussed in Gao and Gijbels (2006), the leading term of each of the existing nonparametric kernel test statistics is of a quadratic form of $\{e_t\}$. The main objective of their paper is to represent the asymptotic distribution of each of such tests by an Edgeworth expansion through using various asymptotic properties of quadratic forms of $\{e_t\}$. The authors have then been able to study large and finite-sample properties of both the size and power functions of such nonparametric kernel tests. Let $K(\cdot)$ be the probability kernel density function and h be the bandwidth involved in the construction of a nonparametric kernel test statistic denoted by $\widehat{L}_T(h)$. In order to implement the kernel test in practice, the authors have also proposed a novel bootstrap simulation procedure to approximate the $1 - \alpha$ quantile of the distribution of the kernel test by a bootstrap simulated critical value l_α. In theory, Gao and Gijbels (2006) have shown that

$$\alpha_T(h) = 1 - \Phi(l_\alpha - S_T) - \kappa_T \left(1 - (l_\alpha - S_T)^2\right) \phi(l_\alpha - S_T)$$
$$+ o\left(\sqrt{h^d}\right), \qquad (3.23)$$

$$\beta_T(h) = 1 - \Phi(l_\alpha - R_T) - \kappa_T \left(1 - (l_\alpha - R_T)^2\right) \phi(l_\alpha - R_T)$$
$$+ o\left(\sqrt{h^d}\right), \qquad (3.24)$$

where $S_T = p_1 \sqrt{h^d}$, $R_T = p_2 T C_T^2 \sqrt{h^d}$, $\kappa_T = p_3 \sqrt{h^d}$, $\Phi(\cdot)$ and $\phi(\cdot)$ denote, respectively, the cumulative distribution and density function of the standard Normal random variable, and all p_i's are positive constants.

To choose a bandwidth $\widehat{h}_{\text{ew}}$ such that $\beta_T(\widehat{h}_{\text{ew}}) = \max_{h \in H_T(\alpha)} \beta_T(h)$ is our objective, where $H_T(\alpha) = \{h : \alpha - c_{\min} < \alpha_T(h) < \alpha + c_{\min}\}$ for some small $c_{\min} > 0$. Gao and Gijbels (2006) have shown that $\widehat{h}_{\text{ew}}$ is proportional to $\left(T C_T^2\right)^{-\frac{3}{2d}}$ when $\Delta(\cdot)$ is a fixed function not depending on T. Such established relationship between C_T and $\widehat{h}_{\text{ew}}$ shows us that the choice of an optimal rate of $\widehat{h}_{\text{ew}}$ depends on that of an order of C_T. For example, the optimal rate of $\widehat{h}_{\text{ew}}$ is proportional to $T^{-\frac{3}{2d}}$ when $\mathcal{H}_{11}$ is a global alternative with $C_T \equiv c$ for some constant c. If C_T is chosen proportional to $T^{-\frac{d+12}{6(d+4)}}$ for a local alternative under $\mathcal{H}_{11}$, then the optimal rate of $\widehat{h}_{\text{ew}}$ is proportional to $T^{-\frac{1}{d+4}}$, which is the order of a nonparametric cross–validation estimation–based bandwidth frequently used for testing purposes. When considering a local alternative with C_T being proportional to $T^{-\frac{1}{2}}\sqrt{\text{loglog}T}$, the optimal rate of $\widehat{h}_{\text{ew}}$ is proportional to $(\text{loglog}T)^{-\frac{3}{2d}}$.

In addition to establishing the main results for (3.2) associated with model (3.1), the authors have then discussed various ways of extending their theory and methodology to optimally choose continuous smoothing parameters in some other testing problems, such as testing for nonparametric significance, additivity and partial linearity, in various nonparametric and semiparametric regression models. The authors have finally mentioned extensions to choose discrete smoothing parameters, such as the number of terms involved in nonparametric series tests and the number of knots in spline–smoothing tests.

3.2.4 An example of implementation

As pointed out in Section 3.1, the implementation of each of existing nonparametric and semiparametric kernel tests involves either a single bandwidth chosen optimally for estimation purposes or a set of bandwidth values. In this section, we show how to implement the proposed test $L_T(h)$ based on $\widehat{h}_{\text{ew}}$ defined in Section 3.2.3 and then compare the finite–sample performance of the proposed choice with that of two alternative versions: (i) the test coupled with a cross–validation bandwidth choice, and (ii) the test associated with an asymptotic critical value.

To assess the finite–sample performance of $\widehat{L}_T(h)$, a normalized version of $L_T(h)$ of (3.12), we consider a nonlinear heteroscedastic time series model of the form

$$Y_t = m(Y_{t-1}) + \sigma(Y_{t-1})\epsilon_t, \ t = 1, \cdots, T, \qquad (3.25)$$

where both $m(\cdot)$ and $\sigma(\cdot) > 0$ are chosen such that $\{Y_t\}$ is strictly stationary, $\{\epsilon_t\}$ is a sequence of independent and identically distributed Normal random errors generated from $N(0,1)$, and $\{Y_s\}$ and $\{\epsilon_t\}$ are mutually independent for all $s \leq t$. We consider a semiparametric model in this example. That is, when specifying $m(\cdot)$ parametrically, we assume that $\sigma(\cdot)$ is already specified parametrically.

Under $\mathcal{H}_{01}$, we generate a sequence of positive resamples $\{Y_t\}$ from

$$Y_t = \alpha_0 + \beta_0 Y_{t-1} + \sigma_0 \sqrt{|Y_{t-1}|}\, \epsilon_t \text{ with } \epsilon_t \sim N(0,1),\ 1 \le t \le T,\ (3.26)$$

where $Y_0 = 0.079$ and the initial parameters are chosen as $\alpha_0 = 0.096$, $\beta_0 = -0.27$ and $\sigma_0 = 0.19$. For such a linear autoregressive model, various probabilistic requirements, such as strict stationarity and mixing conditions, are satisfied automatically.

Under $\mathcal{H}_{11}$, we generate a sequence of positive resamples from

$$Y_t = \alpha_0 + \beta_0 Y_{t-1} + C_T\, \Delta(Y_{t-1}) + \sigma_0 \sqrt{|Y_{t-1}|}\, \epsilon_t,\ 1 \le t \le T,\quad (3.27)$$

where $\Delta(\cdot)$ and C_T are both to be chosen.

In the implementation of $\widehat{L}_T(h)$ as well as its bootstrapping version, we set $\Delta(\cdot) \equiv 1$ and $C_T = T^{-\frac{1}{2}}\sqrt{\mathrm{loglog}(T)}$. In addition, we choose $K(\cdot)$ as the standard normal density function.

The theory and methodology proposed in Gao and Gijbels (2006) suggests using an optimal bandwidth $\widehat{h}_{\mathrm{ew}}$ of the form

$$\widehat{h}_{\mathrm{ew}} = \left(\frac{27}{8\sqrt{\pi}}\right)^{\frac{1}{4}} \left(\frac{\widehat{\sigma}_T^2}{T} \sum_{t=1}^{T} |Y_{t-1}|\right)^{\frac{3}{2}} \cdot (\mathrm{loglog}(T))^{-\frac{3}{2}}, \qquad (3.28)$$

where $\widehat{\sigma}_T^2$ is a $\sqrt{T}$–consistent estimator of σ_0^2.

In order to compare the size and power properties of the proposed test $\widehat{L}_T(h)$ with the most relevant alternatives, we introduce the following simplified notation:

$$\alpha_{10} = P\left(\widehat{L}_T\left(\widehat{h}_{\mathrm{ew}}\right) > l_\alpha^*\left(\widehat{h}_{\mathrm{ew}}\right) \big| \mathcal{H}_{01} \text{ holds}\right),$$

$$\beta_{10} = P\left(\widehat{L}_T\left(\widehat{h}_{\mathrm{ew}}\right) > l_\alpha^*\left(\widehat{h}_{\mathrm{ew}}\right) \big| \mathcal{H}_{11} \text{ holds}\right),$$

$$\alpha_{11} = P\left(\widehat{L}_T\left(\widehat{h}_{\mathrm{cv}}\right) > l_\alpha^*\left(\widehat{h}_{\mathrm{cv}}\right) \big| \mathcal{H}_{01} \text{ holds}\right),$$

$$\beta_{11} = P\left(\widehat{L}_T\left(\widehat{h}_{\mathrm{cv}}\right) > l_\alpha^*\left(\widehat{h}_{\mathrm{cv}}\right) \big| \mathcal{H}_{11} \text{ holds}\right),$$

$$\alpha_{12} = P\left(\widehat{L}_T\left(\widehat{h}_{\mathrm{cv}}\right) > z_\alpha | \mathcal{H}_{01} \text{ holds}\right),$$

$$\beta_{12} = P\left(\widehat{L}_T\left(\widehat{h}_{\mathrm{cv}}\right) > z_\alpha | \mathcal{H}_{11} \text{ holds}\right), \qquad (3.29)$$

where $\widehat{h}_{\mathrm{cv}}$ is given by

$$\widehat{h}_{\mathrm{cv}} = 1.06 \cdot T^{-\frac{1}{5}} \cdot \sqrt{\frac{1}{T-1} \sum_{t=1}^{T} (Y_t - \overline{Y})^2} \text{ with } \overline{Y} = \frac{1}{T} \sum_{t=1}^{T} Y_t,$$

which is an optimal bandwidth based on a cross–validation density estimation method (see Silverman 1986). The reason for choosing $\hat{h}_{cv}$ is that such an estimation-based optimal bandwidth has been commonly used in such autoregressive model specification (Hong and Li 2005).

In the implementation of the simulation procedure, we consider cases where the number of replications of each of the sample versions of α_{1j} and β_{1j} for $j = 0, 1, 2$ was $M = 1000$, each with $B = 250$ number of bootstrapping resamples, and the simulations were done for data sets of sizes $n = 400, 500$ and 600.

In our finite–sample study, we use $z_{0.01} = 2.33$ at the 1% level, $z_{0.05} = 1.645$ at the 5% level and $z_{0.10} = 1.28$ at the 10% level. The detailed results are given in Tables 3.1–3.3 below.

Table 3.1. Rejection rates at the 1% significance level

Sample Size	Null Hypothesis Is True			Null Hypothesis Is False		
T	α_{10}	α_{11}	α_{12}	β_{10}	β_{11}	β_{12}
400	0.014	0.011	0.026	0.134	0.012	0.029
500	0.011	0.016	0.030	0.156	0.015	0.030
600	0.011	0.010	0.024	0.140	0.018	0.034

Table 3.2. Rejection rates at the 5% significance level

Sample Size	Null Hypothesis Is True			Null Hypothesis Is False		
T	α_{10}	α_{11}	α_{12}	β_{10}	β_{11}	β_{12}
400	0.053	0.055	0.071	0.230	0.053	0.066
500	0.063	0.056	0.068	0.246	0.054	0.066
600	0.044	0.048	0.062	0.231	0.057	0.075

Table 3.3. Rejection rates at the 10% significance level

Sample Size	Null Hypothesis Is True			Null Hypothesis Is False		
T	α_{10}	α_{11}	α_{12}	β_{10}	β_{11}	β_{12}
400	0.103	0.112	0.110	0.303	0.096	0.093
500	0.104	0.107	0.105	0.307	0.099	0.098
600	0.096	0.104	0.103	0.291	0.101	0.100

Tables 3.1–3.3 report some comprehensive simulation results for both the

sizes and power values of the proposed tests for models (3.26) and (3.27). In each case, the test based on $\widehat{h}_{\mathrm{ew}}$ has much better properties than the corresponding version based on the CV bandwidth $\widehat{h}_{\mathrm{cv}}$ commonly used in the literature for testing purposes. By comparing the use of a simulated critical value based on the CV optimal bandwidth and the use of an asymptotic critical value in each case, columns 3 and 4 show that there are some severe size distortions particularly when using an asymptotic critical value at the 1% significance level. The corresponding power values are almost comparable with the corresponding sizes as can be seen from columns 6 and 7 in Table 3.1.

While at both the 5% and 10% significance levels there is much less size distortion, the corresponding power values become quite different. For example, at the 5% level, Table 3.2 shows that in each case the test based on $\widehat{h}_{\mathrm{ew}}$ in column 5 is much more powerful than either that based on the CV bandwidth $\widehat{h}_{\mathrm{cv}}$ in column 6 or the test associated with the use of an asymptotic critical value in column 7. Similar features are observable for the 10% case. This further supports our view that the use of an estimation–based optimal bandwidth is not optimal for testing purposes. In addition, the use of an asymptotical normal test is not practically applicable particularly when the sample size is not sufficiently large.

We finally would like to stress that the proposed tests based on the power–optimal bandwidths have not only stable sizes even at a medium sample size of $T \leq 600$, but also reasonable power values even when the "distance" between the null and the alternative has been deliberately close at the rate of $\sqrt{T^{-1} \log\log(T)} = 0.0556$ for $T = 600$ for example. We can expect that the tests would have bigger power values when the "distance" is made wider. Overall, Tables 3.1–3.3 show that the established theory and methodology is implementable and workable in the finite–sample case.

Remark 3.1. The paper by Gao and Gijbels (2006) has addressed the issue of how to appropriately choose bandwidth parameters when using nonparametric and semiparametric kernel–based tests. Both the size and power properties of such tests have been studied. The established theory and methodology has shown that an appropriate bandwidth can be optimally chosen after appropriately balancing the size and power functions. Furthermore, the novel methodology has resulted in an explicit representation for such an optimal bandwidth in the finite–sample case.

The finite–sample studies show that the use of an asymptotically normal test associated with an estimation based optimal bandwidth may not make such a test practically applicable due to poor size and power properties. However, the performance of such a test can be significantly improved when it is coupled with a power–based optimal bandwidth as well as a bootstrap simulated critical value.

3.3 Testing for semiparametric variance models

Specification testing of conditional variance functions is particularly relevant when the conditional mean function is already specified, and the interest is on the conditional variance function. Recent papers such as Dette (2002), and Casas and Gao (2005), among others, are concerned, for example, with testing for a constant conditional variance function.

Consider a semiparametric heteroscedastic model of the form

$$Y_t = m_{\theta_0}(X_t) + e_t = m_{\theta_0}(X_t) + \sigma(X_t)\epsilon_t, \tag{3.30}$$

where $m_{\theta_0}(\cdot)$ is a known parametric function indexed by θ_0, a vector of unknown parameters, $\sigma(\cdot) > 0$ may be an unknown function, and $\{\epsilon_t\}$ is a sequence of i.i.d. errors with $E[\epsilon_t] = 0$ and $E[\epsilon_t^2] = 1$.

We are now interested in testing

$$\mathcal{H}_{02}: \sigma^2(x) = \sigma_{\vartheta_0}^2(x) \text{ versus } \mathcal{H}_{12}: \sigma^2(x) = \sigma_{\vartheta_1}^2(x) + C_{2T}\Delta_2(x), \tag{3.31}$$

for all $x \in \mathbf{R}^d$ and some ϑ_0 and ϑ_1, where ϑ_0 may be different from θ_0, C_{2T} and $\Delta_2(\cdot)$ are defined similarly to C_T and $\Delta(\cdot)$. The key difference here is that we need to choose suitable C_{2T} and $\Delta_2(x)$ such that $\sigma^2(x)$ is positive uniformly in $x \in \mathbf{R}^d$.

Before proposing a test statistic for (3.31), we rewrite model (3.30) into a notational version of the form under $\mathcal{H}_{02}$,

$$\begin{aligned} Y_t &= m_{\theta_0}(X_t) + e_t \text{ with } e_t = \sigma_{\vartheta_0}(X_t)\epsilon_t, \\ e_t^2 &= \sigma_{\vartheta_0}^2(X_t) + \eta_t \text{ with } \eta_t = \sigma_{\vartheta_0}^2(X_t)(\epsilon_t^2 - 1), \end{aligned} \tag{3.32}$$

where $\{\eta_t\}$ satisfies $E[\eta_t] = 0$ and $E[\eta_t^2] = E\left[\sigma_{\vartheta_0}^4(X_t)\right] E\left[(\epsilon_t^2 - 1)^2\right]$ under $\mathcal{H}_{02}$.

Equation (3.32) shows that $\sigma_{\vartheta_0}^2(\cdot)$ may be viewed as the conditional mean function of e_t^2, which may be estimated by $\widehat{e}_t^2 = \left(Y_t - m_{\widehat{\theta}}(X_t)\right)^2$. Similarly to the construction of $L_T(h)$, we thus suggest using a kernel–based test of the form

$$L_{0T}(h) = \sum_{s=1}^{T} \sum_{t=1,\neq s}^{T} \widehat{\eta}_s \, L_2\left(\frac{X_s - X_t}{h_2}\right) \widehat{\eta}_t, \tag{3.33}$$

where $\widehat{\eta}_s = \widehat{e}_s^2 - \sigma_{\widehat{\vartheta}}^2(X_s)$, in which $\widehat{\vartheta}$ is a $\sqrt{T}$–consistent estimator of

ϑ_0, $L_2(\cdot)$ is a probability kernel density function and h_2 is a bandwidth parameter.

As an alternative to (3.32), model (3.32) can be written as

$$
\begin{aligned}
\log\left(Y_t - m_{\theta_0}(X_t)\right)^2 &= \log\left(\sigma^2_{\vartheta_0}(X_t)\right) + \log(\epsilon_t^2) \\
&= \mu_\epsilon + \log\left(\sigma^2_{\vartheta_0}(X_t)\right) + \zeta_t, \qquad (3.34)
\end{aligned}
$$

where $\mu_\epsilon = E\left[\log(\epsilon_t^2)\right]$ and $\zeta_t = \log(\epsilon_t^2) - \mu_\epsilon$. Thus, the function $\sigma^2_{\vartheta_0}(X_t)$ is involved as the conditional mean function of model (3.34). A test statistic similar to $L_{0T}(h)$ may be constructed.

When $X_t = Y_{t-1}$, a normalized version of $L_{0T}(h)$ may be applied for diffusion specification in continuous–time models. Existing studies include Aït-Sahalia (1996b), Fan and Zhang (2003), Gao and King (2004), Chen and Gao (2005), and Hong and Li (2005).

Also, simple calculations imply that for sufficiently large T

$$
\text{var}[L_{0T}(h)] = \sigma_T^2\left(1 + o(1)\right), \qquad (3.35)
$$

where $\sigma_T^2 = 2\mu_2^2 \int K^2(u)du$ with $\mu_2 = E[\eta_1^2]$.

For the implementation of $L_{0T}(h)$ in practice, in order to avoid nonparametrically estimating any unknown quantity we estimate σ_T^2 under $\mathcal{H}_{02}$ by $\widehat{\sigma}_T^2 = 2\widehat{\mu}_2^2 \int K^2(u)du$ with

$$
\widehat{\mu}_2 = \frac{2}{T}\sum_{t=1}^{T}\sigma^2_{\widehat{\vartheta}_0}(X_t).
$$

We then propose using a normalized version of the form

$$
\begin{aligned}
\widehat{L}_{0T}(h) &= \frac{\sum_{s=1}^{T}\sum_{t=1,\neq t}^{T}\widehat{\eta}_s\, L_2\left(\frac{X_s - X_t}{h_2}\right)\widehat{\eta}_t}{\widehat{\sigma}_T} \\
&= \frac{\sum_{t=1}^{T}\sum_{s=1,\neq t}^{T}\eta_s\, L_2\left(\frac{X_s - X_t}{h_2}\right)\eta_t}{\sigma_0}\cdot\frac{\sigma_0}{\widehat{\sigma}_T} \\
&\quad + \frac{1}{\widehat{\sigma}_T}\sum_{t=1}^{T}\sum_{s=1,\neq t}^{T}L_2\left(\frac{X_s - X_t}{h_2}\right) \\
&\quad \times \left(m_\theta(X_s) - m_{\widehat{\theta}}(X_s)\right)^2\left(m_\theta(X_t) - m_{\widehat{\theta}}(X_t)\right)^2 \\
&\quad + \frac{1}{\widehat{\sigma}_T}\sum_{t=1}^{T}\sum_{s=1,\neq t}^{T}L_2\left(\frac{X_s - X_t}{h_2}\right) \\
&\quad \times \left(\sigma^2(X_s) - \sigma^2_{\vartheta_0}(X_s)\right)\left(\sigma^2(X_t) - \sigma^2_{\vartheta_0}(X_t)\right)
\end{aligned}
$$

$$+ \quad o_P \left(\widehat{L}_{0T}(h) \right), \tag{3.36}$$

where $\sigma_0^2 = 2\mu_0^2 \int K^2(u) du$ with $\mu_0 = 2E \left[\sigma_{\vartheta_0}^4(X_1) \right]$ under $\mathcal{H}_{02}$.

The following result establishes that $\widehat{L}_{0T}(h)$ is asymptotically normal under $\mathcal{H}_{02}$; its proof is mentioned in Section 3.5 of this chapter.

Theorem 3.3. *Suppose that Assumptions 3.1 and 3.4 listed in Section 3.5 hold. Then under $\mathcal{H}_{02}$*

$$\lim_{T \to \infty} P \left(\widehat{L}_{0T}(h) \le x \right) = \Phi(x). \tag{3.37}$$

As pointed out in the literature, such asymptotically normal tests may not be very useful in practice, in particular when the size of the data is not sufficiently large. Thus, the conventional α–level asymptotic critical value, l_{acv}, of the standard normality may not be useful in applications. Part of the contribution of this chapter is a proposal of approximating l_{acv} by a bootstrap simulated critical value.

Let $l_{0\alpha}$ $(0 < \alpha < 1)$ be the $1 - \alpha$ quantile of the exact finite–sample distribution of $\widehat{L}_{0T}(h)$. Since $l_{0\alpha}$ may be unknown in practice, we suggest approximating $l_{0\alpha}$ by a simulated α–level critical value, $l_{0\alpha}^*$, using the following simulation procedure:

1. For each $t = 1, 2, \ldots, T$, generate $Y_t^* = m_{\widehat{\theta}}(X_t) + \sigma_{\widehat{\vartheta}_0}(X_t) \, e_t^*$, where the original sample $\mathcal{X}_T = (X_1, \cdots, X_T)$ acts in the resampling as a fixed design, $\{e_t^*\}$ is independent of $\{X_t\}$ and sampled identically distributed from $N(0, 1)$.

2. Use the data set $\{(X_t, Y_t^*) : t = 1, 2, \ldots, T\}$ to re-estimate (θ, ϑ_0). Let $(\widehat{\theta}^*, \widehat{\vartheta}_0^*)$ denote the pair of the resulting estimates. Define $\widehat{L}_{0T}^*(h)$ be the version of $\widehat{L}_{0T}(h)$ with (X_t, Y_t) and $(\widehat{\theta}, \widehat{\vartheta}_0)$ being replaced by (X_t, Y_t^*) and $(\widehat{\theta}^*, \widehat{\vartheta}_0^*)$ in the calculation.

3. Repeat the above steps many times and then obtain the empirical distribution of $\widehat{L}_{0T}^*(h)$. The bootstrap distribution of $\widehat{L}_{0T}^*(h)$ given $\mathcal{W}_T = \{(X_t, Y_t) : 1 \le t \le T\}$ is defined by $P^* \left(\widehat{L}_{0T}^*(h) \le x \right) = P \left(\widehat{L}_{0T}^*(h) \le x | \mathcal{W}_T \right)$. Let $l_{0\alpha}^*$ satisfy $P^* \left(\widehat{L}_{0T}^*(h) \ge l_{0\alpha}^* \right) = \alpha$ and then estimate $l_{0\alpha}$ by $l_{0\alpha}^*$.

We then have the following theorem; its proof is mentioned in Section 3.5 of this chapter.

Theorem 3.4. (i) *Suppose that Assumptions 3.1 and 3.4 listed in Section*

3.5 hold. Then under $\mathcal{H}_{02}$ the following equation

$$\sup_{x \in R^1} \left| P^*(\widehat{L}_{0T}^*(h) \leq x) - P(\widehat{L}_{0T} \leq x) \right| = O\left(\sqrt{h^d}\right) \qquad (3.38)$$

holds in probability with respect to the joint distribution of $\mathcal{W}_T$, and under $\mathcal{H}_{02}$

$$P\left(\widehat{L}_{0T}(h) > l_{0\alpha}^*\right) = \alpha + O\left(\sqrt{h^d}\right). \qquad (3.39)$$

(ii) *Assume that Assumptions 3.1, 3.4 and 3.5 listed in Section 3.5 hold. Then under $\mathcal{H}_{12}$*

$$\lim_{T \to \infty} P\left(\widehat{L}_{0T}(h) > l_{0\alpha}^*\right) = 1. \qquad (3.40)$$

The conclusions of Theorem 3.4 are similar to those of Theorem 3.2.

3.4 Testing for other semiparametric models

As we pointed out before, we will need to consider using a dimension reduction procedure when the dimensionality of $\{X_t\}$ is greater than three. In this section, we discuss how to choose a suitable bandwidth parameter when we consider testing for an alternative model. Our alternative forms for $m(\cdot)$ include nonparametric subset regression, partially linear regression, additive regression and semiparametric single–index regression.

3.4.1 Testing for subset regression

Assume that we can write the model as

$$Y_t = m(X_t) + e_t = m(U_t, V_t) + e_t, \qquad (3.41)$$

where $X_t = (U_t^\tau, V_t^\tau)^\tau$, and U_t and V_t are subsets of X_t. In this section, we are interested to test whether the null hypothesis

$$\mathcal{H}_{03}: \ E[Y_t|X_t] - E[Y_t|U_t] = 0 \qquad (3.42)$$

holds almost surely with respect to the distribution of $\{X_t\}$.

As for nonparametric regression estimation, we estimate

$$m_1(U_t) = E[Y_t|U_t] \ \text{ by } \ \widehat{m}_1(U_t) = \sum_{s=1}^{T} w_{1st} Y_s,$$

where $w_{1st} = \dfrac{K_1\left(\frac{U_s - U_t}{h}\right)}{\sum_{v=1}^{T} K_1\left(\frac{U_s - U_v}{h}\right)}$, in which $K_1(\cdot)$ is a probability kernel function.

This suggests using a test statistic of the form

$$L_{1T}(h) = \sum_{s=1}^{T} \sum_{t=1}^{T} \widehat{Y}_s \ K\left(\frac{X_s - X_t}{h}\right) \ \widehat{Y}_t, \tag{3.43}$$

where $\widehat{Y}_s = [Y_s - \widehat{m}_1(U_s)]\widehat{f}_1(U_s)$ with $\widehat{f}_1(U_s) = \frac{1}{Th^{d_u}} \sum_{t=1}^{T} K_1\left(\frac{U_s - U_t}{h}\right)$, in which d_u is the dimensionality of U_i.

Existing results include Fan and Li (1996), Lavergne and Vuong (1996, 2000), Li (1999), Lavergne (2001), Gao and King (2005), and others. Meanwhile, such testing issues have also been treated in the model selection literature. Recent studies include Fan and Li (2001, 2002), González–Manteiga, Quintela–del–Río and Vieu (2002), and others.

3.4.2 Testing for partially linear regression

The interest here is to test

$$\begin{aligned}
\mathcal{H}_{04} : m(x) &= u^\tau \beta + g(v) \quad \text{versus} \\
\mathcal{H}_{14} : m(x) &= u^\tau \beta + g(v) + C_{4T}\Delta_4(x)
\end{aligned} \tag{3.44}$$

for all $x \in R^d$, where C_{4T} and $\Delta_4(\cdot)$ are similar to what has been defined before, and u and v are subvectors of $x = (u^\tau, v^\tau)^\tau$.

Similarly to (3.43), we construct the following test statistic

$$L_{2T}(h) = \sum_{s=1}^{T} \sum_{t=1}^{T} \widehat{Y}_s \ K\left(\frac{X_s - X_t}{h}\right) \ \widehat{Y}_t, \tag{3.45}$$

where

$$\widehat{Y}_s = Y_s - U_s^\tau \widehat{\beta} - \widehat{g}(V_s), \quad \widehat{\beta} = (\widetilde{U}^\tau \widetilde{U})^+ \widetilde{U}^\tau \widetilde{Y}, \quad \widehat{g}(V_s) = \sum_{t=1}^{T} w_{2st}(Y_t - U_t^\tau \widehat{\beta}),$$

in which $\widetilde{U} = (I - W_2)U$, $U = (U_1, \ldots, U_n)^\tau$, $\widetilde{Y} = (I - W_2)Y$, $W_2 = \{w_{2st}\}$ is a $T \times T$ matrix with $w_{2st} = \frac{K_2\left(\frac{V_s - V_t}{h}\right)}{\sum_{u=1}^{T} K_2\left(\frac{V_s - V_u}{h}\right)}$ with $K_2(\cdot)$ being a kernel function. The form of $L_{2T}(h)$ is similar to some existing results, such as those given in Fan and Li (1996), Li (1999), Härdle, Liang and Gao (2000), Gao and King (2005), and others.

3.4.3 Testing for single–index regression

One of the most efficient dimension reduction procedures is semiparametric single–index modeling. We thus look at testing

$$\mathcal{H}_{05} : m(x) = \psi(x^\tau \theta) \quad \text{versus} \quad \mathcal{H}_{15} : m(x) = \psi(x^\tau \theta) + C_{5T}\Delta_5(x) \tag{3.46}$$

for all $x \in \mathbf{R}^d$, where $\psi(\cdot)$ is an unknown function over $\mathbf{R}^1$, and θ is a vector of unknown parameters. For each given θ, estimate $\psi(\cdot)$ by

$$\widehat{\psi}(X_t^\tau \theta) = \sum_{s=1}^{T} w_{3st} Y_s, \qquad (3.47)$$

where $w_{3st} = \dfrac{K_3 \left(\frac{(X_s - X_t)^\tau \theta}{h} \right)}{\sum_{u=1}^{T} K_3 \left(\frac{(X_s - X_u)^\tau \theta}{h} \right)}$ with $K_3(\cdot)$ being a kernel function

defined over $\mathbf{R}^1$. The parameter θ is then estimated by

$$\widehat{\theta} = \arg \min_{\theta \in \Theta} \sum_{t=1}^{T} \left(Y_t - \widehat{\psi}(X_t^\tau \theta) \right)^2, \qquad (3.48)$$

where Θ is chosen such that the true value of θ is identifiable. This suggests using a test statistic of the form

$$L_{3T}(h) = \sum_{s=1}^{T} \sum_{t=1}^{T} \widehat{Y}_s \, K_3 \left(\frac{(X_s - X_t)^\tau \widehat{\theta}}{h} \right) \widehat{Y}_t, \qquad (3.49)$$

where $\widehat{Y}_t = \left(Y_t - \widehat{\psi}(X_t^\tau \widehat{\theta}) \right) \widehat{f}_3(X_t^\tau \widehat{\theta})$, in which

$$\widehat{f}_3(X_s^\tau \widehat{\theta}) = \frac{1}{Th} \sum_{t=1}^{T} K_3 \left(\frac{(X_s - X_t)^\tau \widehat{\theta}}{h} \right).$$

For the hypotheses in (3.46), there are only few results available in the literature, such as those by Fan and Li (1996), Gao and Liang (1997), Li (1999), Stute and Zhu (2005), and others.

3.4.4 Testing for partially linear single–index regression

As a natural extension, we may consider testing

$$\begin{aligned}
\mathcal{H}_{06} : m(x) &= x^\tau \theta + \psi(x^\tau \eta) \text{ versus} \\
\mathcal{H}_{16} : m(x) &= x^\tau \theta + \psi(x^\tau \eta) + C_{6T} \Delta_6(x)
\end{aligned} \qquad (3.50)$$

for all $x \in \mathbf{R}^d$, where both θ and η are vectors of unknown parameters, and $\psi(\cdot)$ is an unknown function. Interest has been on the estimation of both the parameters and the function in the literature, such as Carroll *et al.* (1997), Xia, Tong and Li (1999), and Xia *et al.* (2004). To the best of our knowledge, little has been done on testing the hypotheses in (3.50).

Modifying $L_{3T}(h)$ in (3.49), we propose using a test statistic of the form

$$L_{4T}(h) = \sum_{s=1}^{T} \sum_{t=1}^{T} \widetilde{Y}_s \, K_3 \left(\frac{(X_s - X_t)^\tau \widehat{\eta}}{h} \right) \widetilde{Y}_j, \qquad (3.51)$$

where $\widehat{\theta}$, $\widehat{\eta}$ and $\widehat{\psi}(\cdot)$ are some consistent estimators as constructed in Chapter 2, and

$$\widetilde{Y}_t = \left(Y_t - X_t^\tau \widehat{\theta} - \widehat{\psi}(X_t^\tau \widehat{\eta}) \right) \widehat{f}_4(X_t^\tau \widehat{\eta}),$$

in which $\widehat{f}_4(X_s^\tau \widehat{\eta}) = \frac{1}{Th} \sum_{t=1}^{T} K_3 \left(\frac{(X_s - X_t)^\tau \widehat{\eta}}{h} \right)$.

3.4.5 Testing for additive regression

The interest here is to test whether $m(x)$ can be decomposed into a sum of p one–dimensional functions as follows:

$$\mathcal{H}_{07} : \; m(x) \;\; = \;\; \sum_{j=1}^{d} m_j(x_j) \;\; \text{versus}$$

$$\mathcal{H}_{17} : \; m(x) \;\; = \;\; \sum_{j=1}^{d} m_j(x_j) + C_{7T} \Delta_7(x), \qquad (3.52)$$

for all $x \in \mathbf{R}^d$, where x_j is the j–th element of $x = (x_1, \cdots, x_d)^\tau$.

Assume that each $m_j(\cdot)$ is estimated by $\widehat{m}_j(X_{tj}) = \sum_{s=1}^{T} W_{Ts}(X_{tj}) Y_s$, in which $\{W_{Ts}(\cdot)\}$ is a sequence of weight functions as constructed in Chapter 2.

We then suggest using a test statistic of the form

$$L_{5T}(h) = \sum_{s=1}^{T} \sum_{t=1}^{T} \widehat{Y}_s \; K \left(\frac{X_s - X_t}{h} \right) \; \widehat{Y}_t, \qquad (3.53)$$

where $\widehat{Y}_t = \left(Y_t - \sum_{j=1}^{p} \widehat{m}_j(X_{tj}) \right) N_t$, in which N_t is chosen to eliminate any random denominator involved in the estimation of $\widehat{m}_j(\cdot)$, such as a kind of density estimator used in $L_{iT}(h)$ for $i = 1, 3, 4$. For testing $\mathcal{H}_{07}$ in (3.52), existing results include Gozalo and Linton (2001), Gao, Tong and Wolff (2002b), Sperlich, Tjøstheim and Yang (2002), Gao and King (2005), and others.

It can be shown that the leading term of each of these tests $L_{iT}(h)$ for $1 \leq i \leq 5$ is of a quadratic form. Thus, we will be able to apply the established theory to study the power function of each of the tests for choosing a suitable h. When there are several different bandwidths involved in the construction of the tests, the study of the power function of each of the tests becomes slightly more complicated than that for the case where only one bandwidth h is involved. For example, we will need to calculate partial derivatives of the power function with respect to individual bandwidth variables.

3.5 Technical notes

This section lists the necessary assumptions for the establishment and the proofs of the main results given in this chapter.

3.5.1 Assumptions

Assumption 3.1. (i) Assume that $\{e_t\}$ is a sequence of i.i.d. continuous random errors with $E[e_t] = 0$, $E[e_t^2] = \sigma^2 < \infty$ and $E[e_t^6] < \infty$.

(ii) $\{X_t\}$ is a sequence of strictly stationary time series variables. In addition, we assume that $\{X_t\}$ is α–mixing with the mixing coefficient $\alpha(t)$ defined by

$$\alpha(t) = \sup\{|P(A \cap B) - P(A)P(B)| : A \in \Omega_1^s, B \in \Omega_{s+t}^\infty\} \leq C_\alpha \alpha^t$$

for all $s, t \geq 1$, where $0 < C_\alpha < \infty$ and $0 < \alpha < 1$ are constants, and Ω_i^j denotes the σ–field generated by $\{X_k : i \leq k \leq j\}$.

(iii) We assume that $\{X_s\}$ and $\{e_t\}$ are independent for all $s \leq t$. Let $\pi(\cdot)$ be the marginal density such that $\int \pi^3(x)dx < \infty$, and $\pi_{\tau_1, \tau_2, \cdots, \tau_l}(\cdot)$ be the joint probability density of $(X_{1+\tau_1}, \ldots, X_{1+\tau_l})$ $(1 \leq l \leq 4)$. Assume that $\pi_{\tau_1, \tau_2, \cdots, \tau_l}(\cdot)$ for all $1 \leq l \leq 4$ do exist and are continuous.

(iv) Assume that the univariate kernel function $K(\cdot)$ is a symmetric and bounded probability density function. In addition, we assume the existence of both $K^{(3)}(\cdot)$, the three–time convolution of $K(\cdot)$ with itself, and $K_2^{(2)}(\cdot)$, the two–time convolution of $K^2(\cdot)$ with itself.

(v) The bandwidth parameter h satisfies both

$$\lim_{T \to \infty} h = 0 \quad \text{and} \quad \lim_{T \to \infty} Th^d = \infty.$$

Assumption 3.2. (i) Let $\mathcal{H}_{01}$ be true. Then $\theta_0 \in \Theta$ and

$$\lim_{T \to \infty} P\left(\sqrt{T}\|\widehat{\theta} - \theta_0\| > B_{1L}\right) < \varepsilon_1$$

for any $\varepsilon_1 > 0$ and some $B_{1L} > 0$.

(ii) Let $\mathcal{H}_{11}$ be true. Then there is a $\theta_1 \in \Theta$ such that

$$\lim_{T \to \infty} P\left(\sqrt{T}\|\widehat{\theta} - \theta_1\| > B_{2L}\right) < \varepsilon_2$$

for any $\varepsilon_2 > 0$ and some $B_{2L} > 0$.

(iii) There exist some absolute constants $\varepsilon_3 > 0$, $\varepsilon_4 > 0$, and $0 < B_{3L}, B_{4L} < \infty$ such that the following

$$\lim_{T \to \infty} P\left(\sqrt{T}\|\widehat{\theta}^* - \widehat{\theta}\| > B_{3L}|\mathcal{W}_T\right) < \varepsilon_3$$

holds in probability, where $\widehat{\theta}^*$ is as defined in the Simulation Procedure above Theorem 3.2.

(iv) Let $m_\theta(x)$ be differentiable with respect to θ and $\frac{\partial m_\theta(x)}{\partial \theta}$ be continuous in both x and θ. In addition,

$$0 < E\left[\left\|\frac{\partial m_\theta(X_1)}{\partial \theta}\Big|_{\theta=\theta_0}\right\|^2 \pi(X_1)\right] < \infty,$$

where the notation $E[f(X_1, \theta_0)]$ denotes the conventional expectation for some function $f(\cdot, \theta_0)$ when X_1 is random, and $\|\cdot\|^2$ denotes the Euclidean norm.

Assumption 3.3. Under $\mathcal{H}_{11}$, we have

$$\lim_{T\to\infty} T\sqrt{h^d}\, C_T^2 = \infty \quad \text{and} \quad 0 < E\left[\Delta^2(X_1)\pi(X_1)\right] < \infty.$$

Assumption 3.4. (i) There exist some absolute constants $\varepsilon_1 > 0$ and $0 < A_{1L} < \infty$ such that

$$\lim_{T\to\infty} P\left(\sqrt{T}\|\widehat{\theta} - \theta\| > A_{1L}\right) < \varepsilon_1.$$

(ii) Let $\mathcal{H}_{02}$ be true. Then $\vartheta_0 \in \Theta$ and

$$\lim_{T\to\infty} P\left(\sqrt{T}\|\widehat{\vartheta}_0 - \vartheta_0\| > B_{1L}\right) < \varepsilon_2$$

for any $\varepsilon_2 > 0$ and some $B_{1L} > 0$.

Let $\mathcal{H}_{02}$ be false. Then there is a $\vartheta_1 \in \Theta$ such that

$$\lim_{T\to\infty} P\left(\sqrt{T}\|\widehat{\vartheta}_0 - \vartheta_1\| > B_{2L}\right) < \varepsilon_3$$

for any $\varepsilon_3 > 0$ and some $B_{2L} > 0$.

(iii) There exist some absolute constants $\varepsilon_4 > 0$, $\varepsilon_5 > 0$, and $0 < B_{3L}, B_{4L} < \infty$ such that both

$$\lim_{T\to\infty} P\left(\sqrt{T}\|\widehat{\vartheta}_0^* - \widehat{\vartheta}_0\| > B_{3L}|\mathcal{W}_T\right) < \varepsilon_4 \quad \text{and}$$

$$\lim_{T\to\infty} P\left(\sqrt{T}\|\widehat{\theta}^* - \widehat{\theta}\| > B_{4L}|\mathcal{W}_T\right) < \varepsilon_5$$

hold in probability, where $\widehat{\vartheta}_0^*$ and $\widehat{\theta}^*$ are as defined in the Simulation Procedure above Theorem 3.2.

(iv) Let $m_\theta(x)$ and $\sigma_\vartheta(x)$ be twice differentiable with respect to θ and ϑ, respectively. In addition, the following quantities are assumed to be finite:

$$\widetilde{C}_1(m) = E\left[\left\|\frac{\partial m_\theta(X_1)}{\partial \theta}\right\|^4\right] \quad \text{and} \quad \widetilde{C}_2(\sigma^2) = E\left[\left\|\frac{\partial \sigma_\vartheta^2(X_1)}{\partial \vartheta}\Big|_{\vartheta=\vartheta_0}\right\|^2\right].$$

Assumption 3.5. Let $\lim_{T\to\infty} T\sqrt{h}\ C_{2T}^2 = \infty$ and $0 < C_1(D) = E\left[D_2^2(X_1)\right] < \infty$.

The first three parts of Assumption 3.1 are quite natural in this kind of problem. We believe that the main results of this chapter remain true when $\{e_t\}$ is a sequence of strictly stationary and α–mixing errors. Since dealing with such dependent errors involves much more technicalities, we impose the i.i.d. conditions on $\{e_t\}$ throughout this chapter. Assumption 3.1(iv) is to ensure the existence of quantities associated with $K(\cdot)$. As pointed out throughout this chapter, Assumption 3.1(v) imposes the minimal conditions on h such that the asymptotic normality is the limiting distribution of each of the proposed tests.

Assumption 3.2 is for some technical proofs and derivatives. Many well–known parametric functions and estimators do satisfy Assumption 3.2. Assumption 3.3 imposes some mild conditions to ensure that both classes of global and local alternatives are included. For the global case where $C_T \equiv c$ (constant), the first part of Assumption 3.3 follows from Assumption 3.1(iv). For local alternatives, numerous choices of $C_T \to 0$ satisfy the first part of Assumption 3.3. Assumption 3.4 is similar to Assumption 3.2, and Assumption 3.5 is analogous to Assumption 3.3.

3.5.2 Technical lemmas

It follows from (3.15) that

$$\kappa_T = \frac{\sqrt{h^d}\left(\frac{\mu_3^2 K^2(0)}{Th^d} + \frac{4\mu_2^3\nu_3}{3}K^{(3)}(0)\right)}{\sigma_T^3}. \tag{3.54}$$

We impose the very natural condition that $\lim_{T\to\infty} Th^d = \infty$ and obtain the following approximations:

$$\begin{aligned}
\sigma_T^2 &= (\mu_4 - \mu_2^2)\frac{K^2(0)}{Th^d} + 2\mu_2^2\nu_2\int K^2(u)du \\
&\approx 2\mu_2^2\nu_2\int K^2(u)du \equiv \sigma_0^2, \\
\kappa_T &= \sigma_T^{-3}\sqrt{h^d}\left(\frac{4\mu_2^3\nu_3}{3}K^{(3)}(0) + \mu_3^2\frac{K^2(0)}{Th^d}\right) \\
&\approx \sqrt{h^d}\,\sigma_0^{-3}\cdot\frac{4\mu_2^3\nu_3}{3}K^{(3)}(0), \tag{3.55}
\end{aligned}$$

where the symbol " $\approx$ " means that both sides are asymptotically identical. Let

$$a_1 = \frac{4K^{(3)}(0)\mu_2^3\nu^3}{3\sigma_0^3} = \frac{\sqrt{2}K^{(3)}(0)}{3}\left(\sqrt{\int K^2(u)du}\right)^{-3} c(\pi) \qquad (3.56)$$

with $c(\pi) = \dfrac{\int \pi^3(x)dx}{\left(\sqrt{\int \pi^2(x)dx}\right)^3}$.

It then follows from (3.55) and (3.56) that

$$\kappa_T \approx a_1 \sqrt{h^d} \qquad (3.57)$$

using $\lim_{T\to\infty} Th^d = \infty$.

In order to establish some useful lemmas without including non–essential technicality, we introduce the following simplified notation:

$$a_{st} = \frac{1}{T\sqrt{h^d}\sigma_0}K\left(\frac{X_s - X_t}{h}\right),$$

$$N_T(h) = \sum_{t=1}^{T}\sum_{s=1,\neq t}^{T} e_s\, a_{st}\, e_t, \qquad (3.58)$$

$$\rho(h) = \frac{\sqrt{2}K^{(3)}(0)\int\pi^3(u)du}{3}\left(\sqrt{\int \pi^2(u)du \int K^2(v)dv}\right)^{-3}\sqrt{h^d},$$

where $\sigma_0^2 = 2\mu_2^2\,\nu_2\int K^2(v)dv$ with $\nu_2 = E[\pi^2(X_1)]$ and $\mu_2 = E[e_1^2]$, and $K^{(3)}(\cdot)$ denotes the three–time convolution of $K(\cdot)$ with itself. We now have the following lemma.

Lemma 3.1. *Suppose that the conditions of Theorem 3.2 hold. Then for any h*

$$\sup_{x\in\mathbf{R}^1}\left|P\left(N_T(h)\leq x\right) - \Phi(x) + \rho(h)\left(x^2 - 1\right)\phi(x)\right| = O\left(h^d\right). \qquad (3.59)$$

PROOF: The proof is based on a nontrivial application of Theorem 1.1 of Götze, Tikhomirov and Yurchenko (2004). As the proof is quite general and useful in itself, it is relegated to Theorem A.2 of the appendix at the end of this book.

Recall $N_T(h) = \sum_{t=1}^{T}\sum_{s=1,\neq t}^{T} e_s\, a_{st}\, e_t$ as defined in (3.58) and let

$$\frac{L_T(h)}{\sigma_0} = \frac{h^{\frac{d}{2}}}{T\sigma_0}\sum_{s=1}^{T}\sum_{t=1,\neq i}^{T}\widehat{e}_s\, K_h(X_s - X_t)\,\widehat{e}_t$$

$$= \frac{h^{\frac{d}{2}}}{T\sigma_0}\sum_{s=1}^{T}\sum_{t=1,\neq s}^{T} e_s\, K_h(X_s - X_t)\, e_t$$

$$+\frac{h^{\frac{d}{2}}}{T\sigma_0} \sum_{s=1}^{T} \sum_{t=1,\neq i}^{T} K_h(X_s - X_t)$$

$$\times \left[m(X_s) - m_{\widehat{\theta}}(X_s)\right] \left[m(X_t) - m_{\widehat{\theta}}(X_t)\right]$$

$$+\frac{2h^{\frac{d}{2}}}{T\sigma_0} \sum_{s=1}^{T} \sum_{t=1,\neq s}^{T} e_s \, K_h(X_s - X_t) \left[m(X_t) - m_{\widehat{\theta}}(X_t)\right]$$

$$= N_T(h) + S_T(h) + D_T(h), \tag{3.60}$$

where $N_T(h) = \frac{h^{\frac{d}{2}}}{T\sigma_0} \sum_{s=1}^{T} \sum_{t=1,\neq s}^{T} e_s \, K_h(X_s - X_t) \, e_t$,

$$S_T(h) = \frac{h^{\frac{d}{2}}}{T\sigma_0} \sum_{s=1}^{T} \sum_{t=1,\neq s}^{T} K_h(X_s - X_t) \tag{3.61}$$

$$\times \left[m(X_s) - m_{\widehat{\theta}}(X_s)\right] \left[m(X_t) - m_{\widehat{\theta}}(X_t)\right],$$

$$D_T(h) = \frac{2h^{\frac{d}{2}}}{T\sigma_0} \sum_{s=1}^{T} \sum_{t=1,\neq s}^{T} e_s \, K_h(X_s - X_t) \left[m(X_t) - m_{\widehat{\theta}}(X_t)\right].$$

We then define $N_T^*(h)$, $S_T^*(h)$ and $D_T^*(h)$ as the corresponding versions of $N_T(h)$, $S_T(h)$ and $D_T(h)$ involved in (3.60) with (X_t, Y_t) and $\widehat{\theta}$ being replaced by (X_t, Y_t^*) and $\widehat{\theta}^*$, respectively.

Lemma 3.2. *Suppose that the conditions of Theorem 3.2(i) hold. Then the following*

$$\sup_{x \in \mathbf{R}^1} \left|P^* \left(N_T^*(h) \leq x\right) - \Phi(x) + \rho(h) \, (x^2 - 1) \, \phi(x)\right| = O_P\left(h^d\right) \tag{3.62}$$

holds in probability.

PROOF: Since the proof follows similarly from that of Lemma 3.1 using some conditioning arguments given $\mathcal{W}_T = \{(X_t, Y_t) : 1 \leq t \leq T\}$, we do not wish to repeat the details.

Lemma 3.3. (i) *Suppose that the conditions of Theorem 3.2(ii) hold. Then under $\mathcal{H}_{01}$*

$$E\left[S_T(h)\right] = O\left(\sqrt{h^d}\right) \quad \text{and} \quad E\left[D_T(h)\right] = o\left(\sqrt{h^d}\right). \tag{3.63}$$

(ii) *Suppose that the conditions of Theorem 3.2(ii) hold. Then under $\mathcal{H}_{01}$*

$$E^*\left[S_T^*(h)\right] = O_P\left(\sqrt{h^d}\right) \quad \text{and} \quad E^*\left[D_T^*(h)\right] = o_P\left(\sqrt{h^d}\right) \tag{3.64}$$

in probability with respect to the joint distribution of $\mathcal{W}_T$, where $E^*[\cdot] = E[\cdot|\mathcal{W}_T]$.

(iii) *Suppose that the conditions of Theorem 3.2(i) hold. Then under $\mathcal{H}_{01}$*

$$E\left[S_T(h)\right] - E^*\left[S_T^*(h)\right] \;=\; O_P\left(\sqrt{h^d}\right) \quad and$$

$$E\left[D_T(h)\right] - E^*\left[D_T^*(h)\right] \;=\; o_P\left(\sqrt{h^d}\right) \tag{3.65}$$

in probability with respect to the joint distribution of $\mathcal{W}_T$.

PROOF: As the proofs of (i)–(iii) are quite similar, we need only to prove the first part of (iii). In view of the definition of $\{a_{st}\}$ of (3.58) and (3.61), we have

$$S_T(h) \;=\; \sum_{t=1}^{T}\sum_{s=1,\neq t}^{T}\left(m(X_s) - m_{\widehat{\theta}}(X_s)\right) a_{st}\left(m(X_t) - m_{\widehat{\theta}}(X_t)\right),$$

$$S_T^*(h) \;=\; \sum_{t=1}^{T}\sum_{s=1,\neq t}^{T}\left(m(X_s) - m_{\widehat{\theta}^*}(X_s)\right) a_{st}\left(m(X_t) - m_{\widehat{\theta}^*}(X_t)\right).$$

Ignoring the higher–order terms, it can be shown that the leading term of $S_T^*(h) - S_T(h)$ is represented approximately by

$$S_T^*(h) - S_T(h) \;=\; (1 + o_P(1))\sum_{t=1}^{T}\sum_{s=1,\neq t}^{T} a_{st} \tag{3.66}$$

$$\times\; \left(m_{\widehat{\theta}}(X_s) - m_{\widehat{\theta}^*}(X_s)\right)\left(m_{\widehat{\theta}}(X_t) - m_{\widehat{\theta}^*}(X_t)\right).$$

Using (3.66), Assumption 3.2 and the fact that

$$E[a_{st}] \;=\; \frac{1}{T\sqrt{h^d}\sigma_0}E\left[K\left(\frac{X_s - X_t}{h}\right)\right]$$

$$=\; \frac{\sqrt{h^d}}{T\sigma_0}\int K(u)du \int \pi^2(v)dv$$

$$=\; \frac{\sqrt{h^d}}{T\sigma_0}\int \pi^2(v)dv, \tag{3.67}$$

we can deduce that

$$E[S_T(h)] - E^*[S_T^*(h)] = O_P\left(\sqrt{h^d}\right), \tag{3.68}$$

which completes an outline of the proof.

Lemma 3.4. *Suppose that the conditions of Theorem 3.2(ii) hold. Then under $\mathcal{H}_{11}$*

$$\lim_{T\to\infty} E\left[S_T(h)\right] = \infty \quad and \quad \lim_{T\to\infty}\frac{E\left[D_T(h)\right]}{E\left[S_T(h)\right]} = 0. \tag{3.69}$$

PROOF: In view of the definitions of $S_T(h)$ and $D_T(h)$, we need only to show the first part of (3.69). Observe that for θ_1 defined in Assumption 3.2(ii),

$$
\begin{aligned}
S_T(h) \quad = \quad & \sum_{t=1}^{T} \sum_{s=1,\neq t}^{T} \left(m(X_s) - m_{\widehat{\theta}}(X_s) \right) a_{st} \left(m(X_t) - m_{\widehat{\theta}}(X_t) \right) \\
= \quad & \sum_{t=1}^{T} \sum_{s=1,\neq t}^{T} (m(X_s) - m_{\theta_1}(X_s)) \, a_{st} \, (m(X_t) - m_{\theta_1}(X_t)) \\
+ \quad & \sum_{t=1}^{T} \sum_{s=1,\neq t}^{T} \left(m_{\theta_1}(X_s) - m_{\widehat{\theta}}(X_s) \right) a_{st} \left(m_{\theta_1}(X_t) - m_{\widehat{\theta}}(X_t) \right) \\
+ \quad & o_P\left(S_T(h) \right).
\end{aligned}
\tag{3.70}
$$

In view of (3.70), using Assumption 3.2(ii), in order to prove (3.69) it suffices to show that as $T \to \infty$ and $h \to 0$,

$$
E\left[\sum_{t=1}^{T} \sum_{s=1,\neq t}^{T} (m(X_s) - m_{\theta_1}(X_s)) \, a_{st} \, (m(X_t) - m_{\theta_1}(X_t)) \right] \to \infty.
\tag{3.71}
$$

Simple calculations imply that as $T \to \infty$ and $h \to 0$

$$
\begin{aligned}
& E\left[\sum_{t=1}^{T} \sum_{s=1,\neq t}^{T} (m(X_s) - m_{\theta_1}(X_s)) \, a_{st} \, (m(X_t) - m_{\theta_1}(X_t)) \right] \\
= \quad & C_T^2 \, E\left[\sum_{t=1}^{T} \sum_{s=1,\neq t}^{T} \Delta(X_s) a_{st} \Delta(X_t) \right] \\
= \quad & \sigma_0^{-1}(1 + o(1)) \, C_T^2 \, \sqrt{h^d} \, T \int K(u)du \int \Delta^2(v)\pi^2(v)dv \\
= \quad & \sigma_0^{-1}(1 + o(1)) \, T \, C_T^2 \sqrt{h^d} \int \Delta^2(v)\pi^2(v)dv \to \infty
\end{aligned}
\tag{3.72}
$$

using Assumption 3.3, where σ_0 is as defined in (3.58).

3.5.3 Proof of Theorem 3.1

The proof follows directly from Theorem A.1 in the appendix.

3.5.4 Proof of Theorem 3.2

Proof of (3.19) of Theorem 3.2: Recall from (3.60) and (3.61) that

$$\widehat{L}_T(h) \quad = \quad (N_T(h) + S_T(h) + D_T(h)) \cdot \frac{\sigma_0}{\overline{\sigma}_T}, \qquad (3.73)$$

$$\widehat{L}_T^*(h) \quad = \quad (N_T^*(h) + S_T^*(h) + D_T^*(h)) \cdot \frac{\sigma_0}{\overline{\sigma}_T^*}, \qquad (3.74)$$

where

$$\overline{\sigma}_T^2 = 2\widehat{\mu}_2^2 \widehat{\nu}_2 \int K^2(u) du \quad \text{with} \quad \widehat{\mu}_2 = \frac{1}{T} \sum_{t=1}^{T} \left(Y_t - m_{\widehat{\theta}}(X_t) \right)^2,$$

$$\overline{\sigma}_T^{*2} = 2\widehat{\mu}_2^{*2} \int K^2(u) du \quad \text{with} \quad \widehat{\mu}_2^* = \frac{1}{n} \sum_{t=1}^{T} \left(Y_t - m_{\widehat{\theta}^*}(X_t) \right)^2,$$

and $\widehat{\nu}_2 = \frac{1}{T} \sum_{t=1}^{T} \widehat{\pi}(X_t)$ with $\widehat{\pi}(\cdot)$ being defined as before.

In view of Assumption 3.2 and Lemmas 3.1–3.3, we may ignore any terms with orders higher than $\sqrt{h^d}$ and then consider the following approximations:

$$\widehat{L}_T(h) \quad = \quad N_T(h) + E\left[S_T(h)\right] + o_P(\sqrt{h^d}) \quad \text{and}$$

$$\widehat{L}_T^*(h) \quad = \quad N_T^*(h) + E^*\left[S_T^*(h)\right] + o_P\left(\sqrt{h^d}\right). \qquad (3.75)$$

Let $s(h) = E[S_T(h)]$ and $s^*(h) = E^*\left[S_T^*(h)\right]$. We then apply Lemmas 3.1 and 3.2 to obtain that

$$P\left(\widehat{L}_T(h) \le x\right) \quad = \quad P\left(N_T(h) \le x - s(h) + o_P\left(\sqrt{h^d}\right)\right)$$

$$= \quad \Phi(x - s(h)) - \rho(h)((x - s(h))^2 - 1)$$

$$\times \quad \phi(x - s(h)) + o\left(\sqrt{h^d}\right) \quad \text{and}$$

$$P^*\left(\widehat{L}_T^*(h) \le x\right) \quad = \quad P^*\left(N_T^*(h) \le x - s^*(h) + o_P\left(\sqrt{h^d}\right)\right)$$

$$= \quad \Phi(x - s^*(h)) - \rho(h)((x - s^*(h))^2 - 1)$$

$$\times \quad \phi(x - s^*(h)) + o_P\left(\sqrt{h^d}\right) \qquad (3.76)$$

hold uniformly over $x \in \mathbf{R}^1$.

Theorem 3.2(i) follows consequently from (3.65) and (3.76).

Proof of (3.20) of Theorem 3.2: In view of the definition that

$$P^*\left(\widehat{L}_T^*(h) \ge l_\alpha^*\right) = \alpha$$

and the conclusion from Theorem 3.2(i) that

$$P\left(\widehat{L}_T(h) \geq l_\alpha^*\right) - P^*\left(\widehat{L}_T^*(h) \geq l_\alpha^*\right) = O_P\left(\sqrt{h^d}\right), \qquad (3.77)$$

the proof of

$$P\left(\widehat{T}_n(h) \geq l_\alpha^*\right) = \alpha + O\left(\sqrt{h^d}\right)$$

follows unconditionally from the dominated convergence theorem.

Proof of (3.21) of Theorem 3.2: Theorem 3.2(ii) follows consequently from Lemma 3.4 and equations (3.75)–(3.76).

3.5.5 Proofs of Theorem 3.3 and 3.4

The proof of Theorem 3.3 follows from an application of Theorem A.1 in the appendix. Similarly to the proof of Theorem 3.2, we may prove Theorem 3.4. The detailed proofs are available from Casas and Gao (2005).

3.6 Bibliographical notes

The literature on nonparametric and semiparametric specification is huge. Härdle and Mammen (1993) have developed consistent tests for a parametric specification by employing the kernel estimation technique. Wooldridge (1992), Yatchew (1992), Gozalo (1993), Samarov (1993), Whang and Andrews (1993), Horowitz and Härdle (1994), Hjellvik and Tjøstheim (1995), Fan and Li (1996), Hart (1997), Zheng (1996), Hjellvik, Yao and Tjøstheim (1998), Li and Wang (1998), Dette (1999), Li (1999), Dette and Von Lieres und Wilkau (2001), Kreiss, Neumann and Yao (2002), Chen, Härdle and Li (2003), Fan and Linton (2003), Zhang and Dette (2004), Gao and King (2005), and others have developed kernel–based consistent tests for various parametric or semiparametric forms (partially linear or single-index) versus nonparametric alternatives for either the independent case or the time series case.

In the same period, Eubank and Spiegelman (1990), Eubank and Hart (1992), Shively, Kohn and Ansley (1994), Chen, Liu and Tsay (1995), Hong and White (1995), Jayasuriya (1996), Hart (1997), Eubank (1999), Härdle, Liang and Gao (2000), Gao, Tong and Wolff (2002b), Li, Hsiao and Zinn (2003), and others have studied series–based consistent tests for a parametric regression model versus nonparametric and semiparametric alternatives. Other related studies include Robinson (1989), Aït–Sahalia, Bickel and Stoker (2001), Gozalo and Linton (2001), Sperlich, Tjøstheim and Yang (2002), Fan and Yao (2003), Fan and Zhang (2003), Gao and King (2005), and others.

This chapter has concentrated on the case where there is no structural break (change–point) in the conditional moments of Y_t given $X_t = x$. When there are structural breaks, nonparametric and semiparametric counterparts of existing parametric tests, such as Andrews (1993), King and Shively (1993), Andrews and Ploberger (1994), Stock (1994), Hansen (2000a, 2000b), and Ling and Tong (2005), have been constructed. Recent papers include Delgado and Hidalgo (2000), Grégoire and Hamrouni (2002), Gijbels and Goderniaux (2004), Gao, Gijbels and Van Bellegem (2006), and others.

Our discussion in this chapter has also been focused on the case where $\{X_t\}$ is a stationary time series while $\{e_t\}$ is a sequence of independent random errors. As briefly mentioned in Example 1.8 and discussed in detailed in Gao et al. (2006), the proposed test statistics may be extended to the case where $\{X_t\}$ is a nonstationary time series.

CHAPTER 4

Model Selection in Nonlinear Time Series

4.1 Introduction

One task in modelling nonlinear time series data is to study the structural relationship between the present observation and the history of the data set. Since Tong (1990), which focuses mainly on parametric models, nonparametric techniques have been used extensively to model nonlinear time series data (see Auestad and Tjøstheim 1990; Tjøstheim 1994; Chapter 6 of Fan and Gijbels 1996; Härdle, Lütkepohl and Chen 1997; Gao 1998; Chapter 6 of Härdle, Liang and Gao 2000; Fan and Yao 2003 and the references therein). Although nonparametric techniques appear feasible, there is a serious problem: the curse of dimensionality. For the independent and identically distributed case, this problem has been discussed and illustrated in several monographs and many papers. In order to deal with the curse of dimensionality problem for the time series case, several nonparametric and semiparametric approaches have been discussed in Chapters 2 and 3, including nonparametric time series single–index and projection pursuit modelling and additive nonparametric and semiparametric time series modelling. In addition to such nonparametric and semiparametric approaches to modelling nonlinear time series, variable selection criteria based on nonparametric techniques have also been discussed in the literature.

As mentioned in Section 3.4.1 of Chapter 3, some variable selection issues have been treated as model specification issues in the literature. To examine the similarity and difference between variable selection and model specification, we provide the following example.

Example 4.1: Consider a nonlinear time series model of the form

$$Y_t = m(X_t) + e_t = m(U_t, V_t) + e_t, \qquad (4.1)$$

where $X_t = (U_t^\tau, V_t^\tau)^\tau$ with U_t and V_t being the subsets of X_t. As has been discussed in Section 3.4.1 of Chapter 3, the form of $m(\cdot, \cdot)$ may be specified through testing a null hypothesis. For example, in order to test whether $\{U_t\}$ is a significant subset, existing results from the econometrics literature suggest testing whether the hypothesis

$$P\left(E[Y_t|U_t, V_t] = E[Y_t|U_t]\right) = 1 \qquad (4.2)$$

holds. In the statistics literature, interest has been on selecting an optimum subset, X_{t_c}, of X_t such that

$$P\left(E[Y_t|X_t] = E[Y_t|X_{t_c}]\right) = 1. \qquad (4.3)$$

The main similarity is that both methods focus on finding the true form of the conditional mean function. The main difference is that unlike the model specification method, the variable selection method may be able to treat each of the components of X_t equally without assuming that $\{V_t\}$ is less significant. As a result, the variable selection method may be more expensive computationally than the model specification method.

Similarly, before using model (1.1) in practice, we should consider identifying a suitable pair (β, g) such that

$$P\left(E[Y_t|X_t] = U_t^\tau \beta + g(V_t)\right) = 1. \qquad (4.4)$$

Such identifiability issue has been addressed in Chen and Chen (1991) and Gao, Anh and Wolff (2001). This chapter thus assumes that the semiparametric form is the true form of the conditional mean function, and the main interest then focuses on a semiparametric selection of the optimum subsets of both U_t and V_t.

In theory, we may suggest using any of existing dimension reduction methods to deal with the dimensionality reduction problem. In practice, however, we need to check whether the method used is appropriate for a given set of data before using it. Although partially linear time series modelling may not be capable of reducing the nonparametric time series regression into a sum of one–dimensional nonparametric functions of individual lags, it can reduce the dimensionality significantly for some cases. Moreover, partially linear time series models take the true structure of the time series data into account and avoid neglecting existing information about linearity of the data. This chapter proposes combining semiparametric time series modelling and nonparametric time series variable selection to deal with the dimensionality reduction problem. We assume that a time series data set $\{(Y_t, U_t, V_t) : t \geq 1\}$ satisfies a partially linear time series model of the form

$$Y_t = U_t^\tau \beta + g(V_t) + e_t, \ t = 1, \cdots, T. \qquad (4.5)$$

In model (4.5), the linear time series component is $U_t^\tau \beta$ and $g(V_t)$ is called the nonparametric time series component.

Model (4.5) covers some existing nonlinear time series models. In theory, model (4.5) can be used to overcome the dimensionality problem, but in practice model (4.5) itself may suffer from the curse of dimensionality. Thus, before using model (4.5) we need to determine whether both the linear and nonparametric components are of the smallest possible dimension. For the partially linear model case, the conventional non-parametric cross–validation model selection function simply cannot take the linear component into account but treats each linear regressor as a nonparametric regressor. As a result, this selection may neglect existing information about the linear component and therefore cause a model misspecification problem. Therefore, we need to consider a novel extension of existing parametric and nonparametric cross–validation model selection criteria to the semiparametric time series setting.

This chapter discusses two different variable selection criteria. The first one, as established by Gao and Tong (2004), proposes using a semi-parametric leave–T_v–out cross–validation function (abbreviated as semi-parametric CVT_v model selection function) for the choice of both the parametric and nonparametric regressors, where $T_v > 1$ is a positive integer satisfying $T_v \to \infty$ as the number of observations, $T \to \infty$. The reason for proposing the CVT_v function rather than the conventional CV1 function is that it yields consistency. The proposed semiparametric cross–validation (CV) model selection procedure has the following features:

(i) It provides a general model selection procedure in determining asymptotically whether both the linear time series component and the nonparametric time series component are of the smallest possible dimension. The procedure can select the true form of the linear time series component. Moreover, it can overcome the curse of dimensionality arising from using nonparametric techniques to estimate $g(\cdot)$ in (4.5).

(ii) It extends the leave–T_v–out cross–validation (CVT_v) selection criterion for linear parametric regression (Shao 1993; Zhang 1993) and the leave–one–out cross–validation (CV1) selection criterion (Vieu 1994; Yao and Tong 1994) for purely nonparametric regression to the semiparametric time series setting. As a result, we also extend the conventional nonparametric CV1 function to a kind of nonparametric CVT_v function.

(iii) It is applicable to a wide variety of models, which include additive partially linear models for both the independent and time series cases. As a result, the proposed model selection procedure is capable of selecting the most significant lags for both the parametric and nonparametric components. Both the methodology and theoretical techniques developed in Gao and Tong (2004) can be used to improve statistical model building and forecasting.

The second selection procedure proposed in Dong, Gao and Tong (2006) is the so–called semiparametric penalty function method by incorporating the leave–one–out cross–validation method with the parametric penalty function method. This chapter then discusses a semiparametric consistent selection procedure suitable for the choice of optimum subsets in a partially linear time series model. The semiparametric penalty function method is implemented using the full set of the data, and simulations show that it works well for both small and medium sample sizes.

4.2 Semiparametric cross–validation method

Although concepts like Akaike (Akaike 1973) information criterion (AIC) and maximum likelihood do not carry over to the nonparametric situation in a straightforward fashion, it makes sense to talk about prediction error and cross–validation in the general framework. The equivalence of AIC and CV criterion for the parametric autoregressive model selection was alluded to by Tong (1976) and established by Stone (1977). Since then, many other authors have studied the behavior of the CV criterion in nonparametric regression for both the independent and time series cases.

Before establishing a general framework for the semiparametric time series case, we need to introduce some notation.

Let $A_q = \{1, \ldots, q\}$, $D_p = \{1, \ldots, p\}$, $\mathcal{A}$ denote all nonempty subsets of A_q and $\mathcal{D}$ denote all nonempty subsets of D_p. For any subset $A \in \mathcal{A}$, U_{tA} is defined as a column vector consisting of $\{U_{ti}, i \in A\}$, and β_A is defined as a column vector consisting of $\{\beta_i, i \in A\}$. For any subset $D \in \mathcal{D}$, V_{tD} is a column vector consisting of $\{V_{ti}, i \in D\}$. We use $d_E = |E|$ to denote the cardinality of a set E. Let $\mathcal{A}_1 = \{A : A \in \mathcal{A}$ such that at least one nonzero component of β is not in $\beta_A\}$,

$$
\begin{aligned}
\mathcal{A}_2 &= \{A : A \in \mathcal{A} \text{ such that } \beta_A \text{ contains all nonzero components of } \beta\}, \\
\mathcal{D}_1 &= \{D : D \in \mathcal{D} \text{ such that } E[Y_t|X_{tD}] = E[Y_t|X_t]\}, \\
\mathcal{D}_2 &= \{D : D \in \mathcal{D} \text{ such that } E[U_t^\tau \beta | X_{tD}] = E[U_t^\tau \beta | X_t]\}, \\
\mathcal{B}_1 &= \{(A, D) : A \in \mathcal{A}_2 \text{ and } D \in \mathcal{D}_1 \cap \mathcal{D}_2\}. \quad (4.6)
\end{aligned}
$$

Obviously, the subsets $A \in \mathcal{A}_1$ and $D \in \mathcal{D}_1^c = \mathcal{D} - \mathcal{D}_1$ correspond to incorrect models. When $(A, D) \in \mathcal{B}_1$,

$$
\begin{aligned}
E[Y_t | U_{tA}, V_{tD}] &= (U_{tA} - E[U_{tA}|V_{tD}])^\tau \beta_A + E[Y_t|V_{tD}] \\
&= (U_t - E[U_t|V_t])^\tau \beta + E[Y_t|V_t] \\
&= E[Y_t | U_t, V_t].
\end{aligned}
$$

This implies that the correct models correspond to $(A_0, D_0) \in \mathcal{B}_1$ such that both A_0 and D_0 are of the smallest dimension. In order to ensure the existence and uniqueness of such a pair (A_0, D_0), we need to introduce some conditions.

Assumption 4.1. (i) Assume that

$$\Delta_{A,D} = E\left\{U_{tA} - E[U_{tA}|V_{tD}]\right\}\left\{U_{tA} - E[U_{tA}|V_{tD}]\right\}^{\top}$$

is a positive definite matrix with order $d_D \times d_D$ for each given pair of $A \in \mathcal{A}$ and $D \in \mathcal{D}$.

(ii) Let $\mathcal{B}_0 = \{(A_0, D_0) \in \mathcal{B}_1,$ such that $|A_0| + |D_0| = \min_{(A,D) \in \mathcal{B}_1}[|A| + |D|]\}$. Assume that (A_0, D_0) is the unique element of $\mathcal{B}_0$ and denoted by (A_*, D_*).

Assumption 4.2. Assume that there is a unique pair (β_*, g_*) such that the true and compact version of model (4.17) is defined by

$$Y_t = U_{tA_*}^{\top} \beta_* + g_*(V_{tD_*}) + e_t, \tag{4.7}$$

where $e_t = Y_t - E[Y_t|U_t, X_t]$.

In order to ensure that model (4.7) is identifiable, we need to impose the following condition to exclude the case where ϕ_* is also a linear function in X_{tj} for $j \in D_p - D_*$.

Assumption 4.3. Define $\theta_j(X_{tj}) = E[g_*(X_{tD_*})|X_{tj}]$ for $j \in D_p - D_*$. There exists an absolute constant $M_0 > 0$ such that

$$\min_{j \in D_p - D_*} \min_{\alpha, \beta} E\left[\theta_j(X_{tj}) - \alpha - \beta X_{tj}\right]^2 \geq M_0.$$

Assumption 4.1(i) requires the definite positivity of the matrix even when both U_t and V_t are dependent time series. When U_t and V_t are independent, $\Delta_{A,D} = E\left\{U_{tA} - E[U_{tA}]\right\}\left\{U_{tA} - E[U_{tA}]\right\}^{\top}$. Clearly, $\mathcal{A}_1$ consists of incorrect subsets A, and the subsets in $\mathcal{A}_2$ may be inefficient because of their unnecessarily large sizes. The optimum pair (A_*, D_*) belongs to $\mathcal{B}_0$, that is, both the parametric and nonparametric regressors are of the smallest dimension. Assumption 4.1(ii) postulates both the existence and uniqueness of (A_*, D_*). It might be possible that there exists another pair (A^*, D^*) such that $|A^*| + |D^*| = |A_*| + |D_*|$. This makes our discussion more complicated. Since it is not a likely case in practice, we discard it.

Assumption 4.2 requires the uniqueness of the pair (β_*, g_*). In other words, Assumption 4.2 also implies that if there is another pair (β^*, ϕ^*) such that

$$U_{tA_*}^{\top} \beta_* + g_*(V_{tD_*}) = U_{tA_*}^{\top} \beta^* + g^*(V_{tD_*}) \quad \text{almost surely},$$

then $\beta^* = \beta_*$ and $g^* = g_*$. Thus, Assumption 4.2 guarantees that the true regression function $U_{tA_*}^\tau \beta_* + g_*(V_{tD_*})$ is identifiable, i.e., β_* and g_* are uniquely determined up to a set of measure zero.

Assumption 4.3 is imposed to exclude the case where $g_*(\cdot)$ has a known parametric linear component. This is just for considerations of rigour. Conventionally, the nonparametric component of a partially linear model is always viewed as a completely nonparametric and nonlinear function.

To establish our procedure for estimating A_* and D_*, consider, for each given pair $A \in \mathcal{A}$ and $D \in \mathcal{D}$, a partially linear model of the form

$$Y_t = U_{tA}^\tau \beta_A + g_D(V_{tD}) + e_t(A, D), \qquad (4.8)$$

where $e_t(A, D) = Y_t - E[Y_t | U_{tA}, X_{tD}]$, β_A is as defined before, and $g_D(\cdot)$ is an unknown function on $\mathbf{R}^{|D|}$. The definition of $e_t(A, D)$ implies that $g_D(X_{tD}) = E[Y_t | V_{tD}] - \beta_A^\tau E[U_{tA} | V_{tD}]$.

In order to estimate β_A and $g_D(\cdot)$, we need to introduce some additional notation:

$$
\begin{aligned}
\hat{g}_{1t}(D) &= \sum_{s=1}^T W_D(t, s) Y_s, \quad \hat{g}_{2t}(A, D) = \sum_{s=1}^T W_D(t, s) U_{sA}, \\
Z_t(D) &= Y_t - \hat{g}_{1t}(D), \quad Z(D) = (Z_1(D), \ldots, Z_T(D))^\tau, \\
W_t(A, D) &= U_{tA} - \hat{g}_{2t}(A, D), \\
W(A, D) &= (W_1(A, D), \ldots, W_T(A, D))^\tau, \\
g_1(X_t) &= E[Y_t | X_t], \quad g_2(X_t) = E[U_t | X_t], \\
W_t &= U_t - g_2(V_t), \quad W = (W_1, \ldots, W_T)^\tau, \qquad (4.9)
\end{aligned}
$$

where

$$W_D(t, s) = \frac{K_D((V_{tD} - V_{sD})/h)}{\sum_{l=1}^T K_D((V_{tD} - V_{lD})/h)},$$

in which T is the number of observations, K_D is a multivariate kernel function defined on $\mathbf{R}^{|D|}$, and h is a bandwidth parameter satisfying

$$h \in H_{TD} = \left[a_D T^{-\frac{1}{4+|D|} - c_D}, b_D T^{-\frac{1}{4+|D|} + c_D} \right],$$

where the constants a_D, b_D and c_D satisfy $0 < a_D < b_D < \infty$ and $0 < c_D < \{2(4 + |D|)\}^{-1}$.

Obviously, there are $(2^p - 1) \times (2^q - 1)$ possible pairs for (A, D). The selection of (A, D) is then carried out by using the data $\{(Y_t, U_t, V_t) : t = 1, \ldots, T\}$ satisfying

$$Y_t = U_t^\tau \beta + g(V_t) + e_t$$

as defined in (4.5). Using (4.8) and (4.9), the least squares estimator of

β_A is (see (1.2.2) of Härdle, Liang and Gao 2000)

$$\widehat{\beta}(A, D) = (W(A, D)^\tau W(A, D))^+ W(A, D)^\tau Z(D), \qquad (4.10)$$

where $(\cdot)^+$ is the Moore–Penrose inverse.

Using model (4.8) fitted based on the data $\{(Y_t, U_t, V_t) : t = 1, \ldots, T\}$, the mean squared prediction error is

$$
\begin{aligned}
L_T(A, D) &= \frac{1}{T} \sum_{t=1}^{T} \left[Z_t(D) - W_t(A, D)^\tau \widehat{\beta}(A, D) \right]^2 \\
&= \frac{1}{T} \left(Z(D) - W(A, D)\widehat{\beta}(A, D) \right)^\tau \left(Z(D) - W(A, D)\widehat{\beta}(A, D) \right) \\
&= \frac{1}{T} \mathcal{E}^\tau R(A, D)\mathcal{E} + \frac{1}{T} \widetilde{G}(D)^\tau R(A, D)\widetilde{G}(D) \\
&\quad + \frac{1}{T}(W\beta)^\tau R(A, D)(W\beta) + \Delta_T(A, D), \qquad (4.11)
\end{aligned}
$$

where

$$
\begin{aligned}
\mathcal{E} &= (e_1, \ldots, e_T)^\tau, \\
P(A, D) &= W(A, D) \left(W(A, D)^\tau W(A, D) \right)^+ W(A, D)^\tau, \\
R(A, D) &= I_T - P(A, D), \quad \widetilde{G}(D) = (\widetilde{g}_1(D), \ldots, \widetilde{g}_T(D))^\tau,
\end{aligned}
$$

$\widetilde{g}_t(D) = g_1(X_t) - \widehat{g}_{1t}(D)$, I_T is the identity matrix of order $T \times T$, and the remainder term is

$$
\begin{aligned}
\Delta_T(A, D) &= \frac{2}{T} \mathcal{E}^\tau R(A, D)\widetilde{G}(D) + \frac{2}{T} \widetilde{G}^\tau(D) R(A, D)(W\beta) \\
&\quad + \frac{2}{T} \mathcal{E}^\tau R(A, D)(W\beta).
\end{aligned}
$$

It follows from (4.11) that the overall expected mean squared error is

$$
\begin{aligned}
M_T(A, D) &= E[L_T(A, D)] = \left(1 - \frac{|A|}{T} \right) \sigma_0^2 \\
&\quad + P_T(A, D) + N_T(A, D) + o_p(M_T(A, D)), \quad (4.12)
\end{aligned}
$$

where $\sigma_0^2 = E[e_t^2]$, $P_T(A, D) = \frac{1}{T} E[(W\beta)^\tau R(A, D)(W\beta)]$, and

$$N_T(A, D) = \frac{1}{T} E\left[\widetilde{G}(D)^\tau R(A, D)\widetilde{G}(D) \right].$$

Equations (4.11) and (4.12) not only reflect the error in model selection and estimation, but also motivate us to generalize the cross–validation method proposed by Shao (1993) for the parametric case to the semiparametric time series case. Suppose that we split the data set into two parts, $\{(Y_t, U_t, V_t) : t \in S\}$ and $\{(Y_t, U_t, V_t) : t \in S^c\}$, where S is a subset of $\{1, \ldots, T\}$ containing T_v integers and S^c is its complement containing T_c integers, $T_v + T_c = T$. Model (4.8) is fitted using the data $\{(Y_t, U_t, V_t) : t \in S^c\}$, called the construction data, and the prediction error is assessed using the data $\{(Y_t, U_t, V_t) : t \in S\}$, treated as if they were future values.

Similarly to (4.10), using the construction data we can estimate β_A by

$$\widehat{\beta}_c(A, D) = (W_c(A, D)^\tau W_c(A, D))^+ W_c(A, D)^\tau Z_c(D), \qquad (4.13)$$

where $W_c(A, D) = (W_{i_1,c}(A, D), \ldots, W_{i_{T_c},c}(A, D))^\tau$ and

$$Z_c(D) = (Z_{i_1,c}(D), \ldots, Z_{i_{T_c},c}(D))^\tau,$$

in which for $t \in S^c$ or $t \in S$,

$$W_{t,c}(A, D) \;=\; U_{tA} - \widehat{g}_{2t}^c(A, D), \;\; \widehat{g}_{2t}^c(A, D) = \sum_{s \in S^c} W_D(t, s) U_{sA},$$

$$Z_{t,c}(D) \;=\; Y_t - \widehat{g}_{1t}^c(D), \;\; \widehat{g}_{1t}^c(D) = \sum_{s \in S^c} W_D(t, s) Y_s.$$

For $t \in S$, let $\widehat{Z}_t^c(A, D) = W_{t,c}(A, D)^\tau \widehat{\beta}_c(A, D)$. The average squared prediction error is then defined by

$$
\begin{aligned}
\mathrm{CV}(A, D; h) \;&=\; \mathrm{CV}(A, D; h, T_v) = \mathrm{CV}_S(A, D; h, T_v) \\
&=\; \frac{1}{T_v} \sum_{t \in S} \left(Z_{t,c}(D) - \widehat{Z}_t^c(A, D) \right)^2 w(V_t), \quad (4.14)
\end{aligned}
$$

where the weight function $w(\cdot)$ is employed to trim off some extreme values of V_t involved in the nonparametric estimator of the density function of V_t in the denominator.

The $\mathrm{CV}(A, D; h)$ function is called the semiparametric leave–T_v–out cross–validation function, abbreviated as semiparametric $\mathrm{CV}T_v$ function. As can be seen from its construction, the parametric and nonparametric counterparts are both special cases. In other words, we extend not only the parametric $\mathrm{CV}T_v$ proposed by Shao (1993) but also the conventional nonparametric CV1 function to the semiparametric $\mathrm{CV}T_v$ function. As the semiparametric CV1 is asymptotically inconsistent in the selection of A, we adopt the following Monte Carlo $\mathrm{CV}T_v$ in the selection of (A, D).

Randomly draw a collection $\mathcal{R}$ of n subsets of $\{1, \ldots, T\}$ that have size T_v and select a model by minimizing

$$
\begin{aligned}
\mathrm{MCCV}(A, D; h) \;&=\; \frac{1}{n} \sum_{S \in \mathcal{R}} \mathrm{CV}_S(A, D; h, T_v) \qquad (4.15) \\
&=\; \frac{1}{nT_v} \sum_{S \in \mathcal{R}} \sum_{t \in S} \left(Z_{t,c}(D) - \widehat{Z}_t^c(A, D) \right)^2 w(V_t);
\end{aligned}
$$

we call it the semiparametric MCCV(T_v) function. Let

$$(\widehat{A}, \widehat{D}, \widehat{h}) = \arg \min_{\{A \in \mathcal{A},\ D \in \mathcal{D},\ h \in H^c_{TD}\}} \text{MCCV}(A, D; h), \qquad (4.16)$$

where

$$H^c_{TD} = \left[a_D T_c^{-\frac{1}{4+|D|}-c_D}, b_D T_c^{-\frac{1}{4+|D|}+c_D} \right],$$

in which the constants a_D, b_D and c_D satisfy $0 < a_D < b_D < \infty$ and $0 < c_D < \{2(4+|D|)\}^{-1}$.

We now state the following main result of this section as established by Gao and Tong (2004).

Theorem 4.1. *Assume that Assumptions 4.1–4.3 and 4.5–4.8 listed in Section 4.5 hold. Then*

$$\lim_{T \to \infty} P\left(\widehat{A} = A_*,\ \widehat{D} = D_* \right) = 1 \ \text{ and }$$

$$\frac{\widehat{h}}{h_*} \to_p 1$$

as $T \to \infty$, where $h_* = c_* T_c^{-\frac{1}{4+|D_*|}}$ and c_* is a positive constant.

Theorem 4.1 shows that if a given data set (Y_t, U_t, V_t) satisfies a model of form (4.5), the proposed semiparametric CVT$_v$ selection procedure suggests that we need only to consider the selection of $(2^q - 1) \times (2^p - 1)$ possible models of the form (4.5). When U_t and V_t are independent, we need only to consider the selection of $2^p + 2^q - 2$ possible models. If we choose to use either a purely nonparametric cross–validation selection procedure or the completely parametric CVT$_v$ selection procedure for the selection of an optimum set of (U_t, V_t), we need to consider the selection of $2^{p+q} - 1$ possible models. Consequently, in theory we may cause a model specification problem, since a completely linear model or a purely nonparametric regression model may be either too simple or too general for a given time series data. In practice, the computation when selecting $2^{p+q} - 1$ possible models is more expensive than that when selecting $(2^q - 1) \times (2^p - 1)$ possible models when p and q are large.

Theorem 4.1 covers two important cases. First, we can select the optimum subset of parametric regressors when the nonparametric component of (4.5) is already compact. Second, Theorem 4.1 also provides a consistent selection procedure for the nonparametric component even when the parametric component is compact. Moreover, the conclusions of Theorem 4.1 apply to many important special cases. For example, the case where U_t, V_t and e_t are all strictly stationary time series is included. This implies that both the proposed model selection procedure and the conclusion of Theorem 4.1 apply to a wide variety of models of the form (4.5).

In addition, Theorem 4.1 not only extends the model selection method of Shao (1993) for the fixed design linear model case to the selection of both parametric and nonparametric regressors in semiparametric time series regression, but also generalizes the conventional nonparametric CV1 function (see Vieu 1994, 1995; Yao and Tong 1994) for both the independent and β–mixing time series cases to the semiparametric $\mathrm{MCCV}(T_v)$ function for the α–mixing time series case.

Theorem 4.1 not only provides the asymptotic consistency of the semiparametric CVT_v selection procedure, but also shows that if a model of form (4.5) within the context tried is the truth, then the semiparametric selection procedure will find it asymptotically.

Before we prove Theorem 4.1 in Section 4.5 below, we discuss some applications of Theorem 4.1 through using Examples 4.2 and 4.3 in Section 4.4 below.

4.3 Semiparametric penalty function method

Section 4.2 has discussed the proposed semiparametric cross–validation selection method. As observed in the simulations in both Shao (1993) and Section 4.2 of this chapter, the number of observations used to fit the model is, however, quite small (with $T_c = 15$ in Shao 1993 for the parametric case and $T_c = 69$ in Section 4.2 for the semiparametric case), while the number of observations used to validate the proposed method is relatively large (with $T_v = 25$ and $T_v = 219$, respectively). This may impede the implementation of the method in practice because the theory requires $T_c \to \infty$; it is more appropriate to use comparatively more data to construct the model but comparatively fewer data to validate the model. In addition to addressing the problem of inconsistency in model selection such as AIC, Zheng and Loh (1995, 1997) have proposed the so–called penalty function method for parametric model selection.

In this section, we propose a semiparametric penalty function-based model selection criterion by incorporating some essential features of the CV1 selection method for the choice of both the parametric and the nonparametric regressors in model (4.5). The main objective of this section is to propose a new selection criterion, establish the associated theory and demonstrate the key feature of easy implementation of the proposed semiparametric penalty function method by using two simulated examples.

Using Equations (4.8)–(4.10), the residual sum of squares is defined by

$$
\begin{aligned}
RSS(A, D; h) &= \sum_{t=1}^{T} \left[Z_t(D) - W_t(A, D)^\tau \widehat{\beta}(A, D) \right]^2 \\
&= \left(Z(D) - W(A, D)\widehat{\beta}(A, D) \right)^\tau \\
&\quad \cdot \left(Z(D) - W(A, D)\widehat{\beta}(A, D) \right) \\
&= \mathcal{E}^\tau R(A, D)\mathcal{E} + \widetilde{G}(D)^\tau R(A, D)\widetilde{G}(D) \\
&\quad + (W\beta)^\tau R(A, D)(W\beta) + T\, \Delta_T(A, D), \quad (4.17)
\end{aligned}
$$

where the quantities involved are defined as in (4.11).

It may be shown from (4.17) that the following equations hold uniformly in $h \in H_{TD}$,

$$
\begin{aligned}
RSS(A, D; h) &= \mathcal{E}^\tau R(A, D)\mathcal{E} + \widetilde{G}(D)^\tau R(A, D)\widetilde{G}(D) \\
&\quad + (W\beta)^\tau R(A, D)(W\beta) + o_P\left(RSS(A, D; h)\right), \\
M_T(A, D; h) &= E\left[RSS(A, D; h)\right] = (T - |A|)\sigma^2 + P_T(A, D) \\
&\quad + N_T(A, D) + o\left(M_T(A, D)\right), \quad (4.18)
\end{aligned}
$$

where

$$
\begin{aligned}
P_T(A, D) &= E\left[(W\beta)^\tau R(A, D)(W\beta)\right] \text{ and} \\
N_T(A, D) &= E\left[\widetilde{G}(D)^\tau R(A, D)\widetilde{G}(D)\right].
\end{aligned}
$$

Before we define our selection procedure, we need to introduce the following penalty function. The penalty function $\Lambda_T(A, D)$ is defined as

$$
\Lambda_T(A, D) : \mathcal{A} \times \mathcal{D} \to \mathbf{R}^1
$$

satisfying the following assumption.

Assumption 4.4. (i) Let $\Lambda_T(\emptyset, \emptyset) = 0$ and $\Lambda_T(\emptyset, D) = 0$ for any given $D \in \mathcal{D}$.

(ii) For any subsets $A_1, A_2 \in \mathcal{A}, D_1, D_2 \in \mathcal{D}$ satisfying $A_2 \supset A_1, D_2 \supset D_1$,

$$
\lim_{T \to \infty} \inf \frac{\Lambda_T(A_1, D_1)}{\Lambda_T(A_2, D_2)} < 1.
$$

(iii) For any nonempty sets A and D, we assume that

$$
\lim_{T \to \infty} \Lambda_T(A, D) = \infty \text{ and } \lim_{T \to \infty} \frac{\Lambda_T(A, D)}{T} = 0.
$$

It should be noted that $\Lambda_T(A, D)$ can be chosen quite generally. $\Lambda_T(A, D)$

is defined as a function of sets in theory, but in practice we can define the penalty function $\Lambda_T(A, D)$ as a function of $|A|$ and $|D|$ satisfying Assumption 4.4. Obviously, the definition of $\Lambda_T(A, D)$ generalizes the function $h_n(k)$ in Zheng and Loh (1997). Assumption 4.4 regularizes the penalty function so as to avoid any problem of either over–fitting or under–fitting.

We now extend the penalty function method from linear model selection to the partially linear model selection. Define

$$(\widetilde{A}, \widetilde{D}, \widetilde{h}) = \arg \min_{A \in \mathcal{A}, D \in \mathcal{D}, h \in H_{TD}} \left\{ RSS(A, D; h) + \Lambda_T(A, D)\widehat{\sigma}^2 \right\},$$
(4.19)

where $\widehat{\sigma}^2 = \frac{1}{T-p}RSS(A_q, D_p; h)$ is the usual consistent estimate of $\text{var}[e_t] = \sigma^2$. It may also be shown from equation (14) of Zheng and Loh (1997) that

$$\widehat{\sigma}^2 = \sigma^2 + o_P(1)$$
(4.20)

uniformly in $h \in H_{TD}$.

The method discussed here generalizes those criteria proposed in Zheng and Loh (1995, 1997), Yao and Tong (1994), and Vieu (1994). For example, if D_* is already identified, then the problem will become a model selection problem for linear models as discussed in Zheng and Loh (1995, 1997). If A is already identified as A_*, and we need only to select D for (A_*, D), then the model selection will reduce to a purely nonparametric leave–one–out cross–validation selection problem. This is because, as shown in Section 2.1 of Gao and Tong (2005), the leading term $\frac{1}{T}RSS(A_*, D; h)$ is asymptotically equivalent to a $CV1(D, h)$ function of (D, h) as defined below:

$$CV1(D, h) = \frac{1}{T} \sum_{t=1}^{T} \left\{ Y_t - U_t^\tau \widehat{\beta}(A_q, D) - \widehat{g}_t(X_{tD}, \widehat{\beta}(A_q, D)) \right\}^2, \quad (4.21)$$

where $\widehat{g}_t(X_{tD}, \beta)$ is as defined before. Thus, in the case where A is already identified as A_*, we may choose (D, h) as follows:

$$(\widetilde{D}, \widetilde{h}) = \arg \min_{D \in \mathcal{D}, h \in H_{TD}} CV1(D, h).$$

We now state the asymptotic consistency of the proposed selection criterion in Theorem 4.2, which is comparable with Theorem 4.1.

Theorem 4.2. *If the Assumptions 4.1–4.4 and 4.5–4.8 listed in Section 4.5 hold, then*

$$\lim_{T \to \infty} P\left(\widetilde{A} = A_*, \widetilde{D} = D_*\right) = 1 \quad and$$

$$\frac{\widetilde{h}}{h_*} \to_P 1$$

as $T \to \infty$, where $h_* = c_* T^{-\frac{1}{4+|D_*|}}$ and c_* is a positive constant.

Before we prove Theorem 4.2 in Section 4.5 below, we examine some finite–sample properties of the proposed penalty function selection criterion through Examples 4.4 and 4.5 in Section 4.4 below.

4.4 Examples and applications

This section illustrates Theorem 4.1 using a simulated model in Example 4.2 and a set of real data in Example 4.3. Theorem 4.2 is then illustrated through using two simulated examples in Examples 4.4 and 4.5.

Example 4.2: Consider a nonlinear time series model of the form

$$
\begin{aligned}
Y_t &= 0.47U_{t-1} - 0.45U_{t-2} + \frac{0.5V_{t-1} - 0.23V_{t-2}}{1 + V_{t-1}^2 + V_{t-2}^2} + e_t, \\
U_t &= 0.55U_{t-1} - 0.12U_{t-2} + \delta_t \quad \text{and} \\
V_t &= 0.3\sin(2\pi V_{t-1}) + 0.2\cos(2\pi V_{t-2}) + \epsilon_t, \quad t = 3, \dots, T, (4.22)
\end{aligned}
$$

where δ_t, ϵ_t and e_t are mutually independent and identically distributed with uniform distributions on $(-1, 1)$, $(-0.5, 0.5)$ and the standard Normal distribution $N(0, 1)$, respectively, U_1, U_2, V_1, V_2 are independent and identically distributed with uniform distribution on $(-1, 1)$, U_s and V_t are mutually independent for all $s, t \geq 3$, and the process $\{(\eta_t, \epsilon_t, e_t)\}$ is independent of both (U_1, U_2) and (V_1, V_2).

For Example 4.2, the strict stationarity and mixing condition can be justified by using existing results (Masry and Tjøstheim 1995, 1997). Thus, Assumption 4.4 holds. For an application of Theorem 4.1, let

$$
\begin{aligned}
\beta &= (\beta_1, \beta_2)^\tau = (0.47, -0.45)^\tau \quad \text{and} \\
g(V_{t-1}, V_{t-2}) &= \frac{0.5V_{t-1} - 0.23V_{t-2}}{1 + V_{t-1}^2 + V_{t-2}^2}.
\end{aligned}
$$

In this example, we consider the case where V_t, V_{t-1} and V_{t-2} are selected as candidate nonparametric regressors and U_{t-1} and U_{t-2} as candidate parametric regressors and then use the proposed semiparametric $\text{MCCV}(T_v)$ function to check if $(U_{t-1}, U_{t-2}, V_{t-1}, V_{t-2})$ is the true semiparametric set. For this case, there are $2^3 - 1 = 7$ possible nonparametric regressors and $2^2 - 1 = 3$ possible parametric regressors. Therefore, there are 10 possible candidates for the true model, since U_t and V_t are independent.

Let $D_0 = \{1, 2\}$, $D_1 = \{0, 1\}$, $D_2 = \{0, 2\}$, $D_3 = \{0, 1, 2\}$, $D_4 = \{0\}$, $D_5 = \{1\}$, $D_6 = \{2\}$, $\mathcal{D} = \{D_i : 0 \le i \le 6\}$, $V_{tD_0} = (V_{t-1}, V_{t-2})^\tau$, $V_{tD_1} = (V_t, V_{t-1})^\tau$, $V_{tD_2} = (V_t, V_{t-2})^\tau$, $V_{tD_3} = (V_t, V_{t-1}, V_{t-2})^\tau$, $V_{tD_4} = V_t$, $V_{tD_5} = V_{t-1}$, $V_{tD_6} = V_{t-2}$, $A_0 = \{1, 2\}$, $A_1 = \{1\}$, $A_2 = \{2\}$, $\mathcal{A} = \{A_i : i = 0, 1, 2\}$, $U_{tA_0} = (U_{t-1}, U_{t-2})^\tau$, $U_{tA_1} = U_{t-1}$ and $U_{tA_2} = U_{t-2}$. It follows that both $D_* = D_0$ and $A_* = A_0$ are unique. Assumptions 4.1–4.3 therefore hold.

Throughout Example 4.2, we use $h \in H_{TD}^c = \left[0.1 \cdot T_c^{-\frac{2}{9}}, 3 \cdot T_c^{-\frac{1}{9}}\right]$, where T_c is to be chosen and the weight function $w(x) = I_{[-1,1]}(x)$, in which $I_A(x)$ is the indicator function. For the multivariate kernel function $K(\cdot)$ involved in $W_D(t, s)$, define $K(u_1, \ldots, u_j) = \prod_{i=1}^{j} k(u_i)$ for $j = 1, \cdots, 3$, where $k(u) = \frac{1}{\sqrt{2\pi}} e^{-\frac{u^2}{2}}$. It follows that Assumptions 4.5–4.7 are all satisfied.

In the calculation of MCCV(T_v), we choose $n = T$, $T_v = T - T_c$ and $T_c = \left[T^{3/4}\right]$, the largest integer part of $T^{3/4}$. Assumption 4.8(ii) follows immediately from the choice of $n = T$ and T_v. Before checking Assumption 4.8(i), we introduce the following notation. Using the independence between U_t and V_t, we have $E[U_{tA_i} | X_{tD_j}] = E[U_{tA_i}]$ for all $i = 0, 1, 2$ and $j = 0, \cdots, 6$. Thus, we need only to introduce the following notation. For $i = 0, 1, 2$, let

$$
\begin{aligned}
\eta_t(A_i) &= U_{tA_i} - E[U_{tA_i}], \\
\eta(A_i) &= (\eta_1(A_i), \ldots, \eta_T(A_i))^\tau, \\
\eta_t &= \eta_t(A_0), \quad \eta = (\eta_1, \ldots, \eta_T)^\tau.
\end{aligned}
$$

Let $\alpha_t = \sum_{j=1}^{2} \eta_t(A_j)\beta_j$ and $\alpha = \eta\beta = (\alpha_1, \ldots, \alpha_T)^\tau$. A detailed calculation yields that

$$
(\eta\beta)^\tau \left(I_T - \eta(A_i)(\eta(A_i)^\tau \eta(A_i))^{-1}\eta(A_i)^\tau\right)(\eta\beta)
$$

$$
= \frac{\sum_{t=3}^{T} \eta_t^2(A_i) \sum_{t=3}^{T} \alpha_t^2 - \left[\sum_{t=3}^{T} \eta_t(A_i)\alpha_t\right]^2}{\sum_{t=3}^{T} \eta_t^2(A_i)} > 0
$$

with probability one for all $i = 1, 2$, because $P(\eta_t(A_i) = \alpha_t) = 0$. This shows that Assumption 4.8(i) holds. Therefore, Assumptions 4.1–4.3 and 4.5–4.8 all hold.

For the three sample sizes $T = 72, 152$ and 302, we calculated the relative frequencies of the selected parametric and nonparametric regressors in 1000 replications. Table 4.1 below reports the results of the simulation for the semiparametric MCCV(T_v) selection function.

Table 4.1. The semiparametric MCCV(T_v) function-based relative
frequencies for Example 4.2

Significant Semiparametric Regressor	Relative Frequency		
	$T = 72$	$T = 152$	$T = 302$
$\{U_{t-1}, U_{t-2}, V_{t-1}, V_{t-2}\}$	0.723	0.891	0.984
$\{U_{t-1}, U_{t-2}, V_t, V_{t-1}, V_{t-2}\}$	0.124	0.064	0.011
$\{U_{t-1}, U_{t-2}, V_{t-1}, V_t\}$	0.065	0.018	0.003
$\{U_{t-1}, U_{t-2}, V_{t-2}, V_t\}$	0.063	0.018	0.002
$\{U_{t-1}, X_{t-1}, V_{t-2}\}$	0.012	0.005	0.000
$\{U_{t-2}, X_{t-1}, V_{t-2}\}$	0.013	0.004	0.000

We now compare the proposed semiparametric MCCV(T_v) selection
function with the conventional nonparametric CV1 selection function.
For the same Example 4.2, consider the case where U_{t-1}, U_{t-2}, V_t, V_{t-1}
and V_{t-2} are selected as candidate nonparametric regressors. For this
case, there are $2^5 - 1 = 31$ possible nonparametric regressors, since we
treat each parametric regressor as nonparametric. As the conventional
nonparametric CV1 function is already consistent, we considered using it
as an alternative to the semiparametric MCCV(T_v) function. To ensure
that the numerical comparison between the semiparametric MCCV(T_v)
method and the nonparametric CV1 model selection can be done in a
reasonable way, we choose the same w, k and h as above and define
$K(u_1, \ldots, u_j) = \prod_{i=1}^{j} k(u_i)$ for $j = 1, \cdots, 5$ for the multivariate kernel
function involved in $W_D(t, s)$. The results based on 1000 replications are
given in Table 4.2.

Table 4.2. The nonparametric CV1-based relative frequencies for
Example 4.2

Significant Nonparametric Regressor	Relative Frequency		
	$T = 72$	$T = 152$	$T = 302$
$\{U_{t-2}, V_{t-1}, V_{t-2}\}$	0.473	0.472	0.477
$\{U_{t-1}, V_{t-1}, V_{t-2}\}$	0.464	0.470	0.476
$\{U_{t-1}, U_{t-2}, V_{t-1}, V_{t-2}\}$	0.016	0.018	0.016
$\{U_{t-1}, U_{t-2}, V_t, V_{t-2}\}$	0.017	0.017	0.016
$\{U_{t-1}, U_{t-2}, V_t, V_{t-1}\}$	0.015	0.019	0.014
$\{V_{t-1}, V_{t-2}\}$	0.013	0.004	0.001
$\{V_t, V_{t-1}, V_{t-2}\}$	0.002	0.000	0.000

Table 4.1 shows that MCCV(T_v) can be implemented in practice and supports the validity of our definition of optimum subset (see Assumption 4.1). The simulation results show that both $\widehat{A}$ and $\widehat{D}$ can be reasonably good estimators of A_* and D_* even when the sample size T is modest. Unlike the semiparametric MCCV(T_v) function, the conventional nonparametric CV1 function cannot identify the true set of regressors $(U_{t-1}, U_{t-2}, V_{t-1}, V_{t-2}\}$. This is a reflection of the fact that the semiparametric MCCV(T_v) selection takes into account the existence of both the parametric and nonparametric regressors while the nonparametric CV1 neglects the existence of the parametric component but treats each parametric regressor as nonparametric. Both the semiparametric and nonparametric CV1 selection procedures considered all possible 10 models for the semiparametric case and all possible 31 models for the nonparametric case. As many other insignificant regressors had zero probability being selected, Tables 4.1 and 4.2 provide only the relative frequencies for the significant regressors. We also recorded the relative frequencies based on the corresponding nonparametric MCCV(T_v). As the values were comparable with those given in Table 4.2, we only report the relative frequencies based on the conventional nonparametric CV1 function. The computation times for the semiparametric selection were much shorter than those for the nonparametric selection.

Throughout Example 4.2, we point out that Assumptions 4.1–4.3 and 4.5–4.8 are all satisfied. In theory, Assumption 4.8(i) is a very minimal model identifiability condition. But, it is not easy to verify in practice. For Example 4.2, however, we have been able to verify the condition.

Example 4.3: Fisheries Western Australia (WA) manages commercial fishing in Western Australia. Simple Catch and Effort statistics are often used in regulating the amount of fish that can be caught and the number of boats that are licensed to catch them. The establishment of the relationship between the Catch (in kilograms) and Effort (the number of days the fishing vessels spent at sea) is very important both commercially and ecologically. This example considers using our model selection procedure to identify a best possible model for the relationship between catch and effort.

The monthly fishing data set from January 1976 to December 1999 is available from the Fisheries WA Catch and Effort Statistics (CAES) database. Existing studies suggest that the relationship between catch and effort is nonlinear while the dependence of the current catch on the past catch appears to be linear. This suggests using a partially linear model of form

$$C_t = \beta_1 C_{t-1} + \cdots + \beta_p C_{t-q} + \phi(E_{t-1}, E_{t-2}, \ldots, E_{t-p}) + \epsilon_t, \ t = r, \cdots,$$

where $r = \max(p, q)$, $\{\epsilon_t\}$ is a random error, and $\{C_t\}$ and $\{E_t\}$ represent the catch and the effort at time t. In computation, we use the transformed data $Y_t = \log_{10} C_t$ and $X_t = \log_{10} E_t$ satisfying the following model

$$Y_{t+r} = \beta_1 Y_{t+r-1} + \ldots + \beta_q Y_{t+r-q} + \phi(X_{t+r-1}, \ldots, X_{t+r-p}) + e_t, \quad t \geq 1,$$
(4.23)

where $\{e_t\}$ is a sequence of strictly stationary error processes with zero mean and finite variance.

Before using model (4.23), we need to choose an optimum and compact form of it. We consider $q = 4$ and $p = 5$ and then find an optimum model; there are $2^4 - 1 = 15$ different parametric and $2^5 - 1 = 31$ different nonparametric regressors.

Similarly to Example 4.2, we define the parametric candidates U_{tA_i} for $1 \leq i \leq 15$ and the nonparametric candidates V_{tD_j} for $1 \leq j \leq 31$. It follows that

$$Y_{t+5} = U_{tA_i}^{\mathsf{T}} \beta_{A_i} + \phi_{D_j}(V_{tD_j}) + e_{tij}, \qquad (4.24)$$

where β_{A_i} and ϕ_{D_j} are similar to those of β_A and ϕ_D, and $\{e_{tij}\}$ is allowed to be a sequence of strictly stationary error processes with zero mean and finite variance.

We then use the 288 observations of the data from January 1976 to December 1999 to select a best possible partially linear model. In the calculation of the $\mathrm{MCCV}(T_v)$ function, we choose $n = T = 288$, $T_c = [T^{3/4}] = 69$ and $T_v = T - T_c = 219$. For this example, we consider using $K(u_1, \ldots, u_j) = \prod_{i=1}^{j} k(u_i)$ for $1 \leq j \leq 9$ for the multivariate kernel function involved in $W_D(t, s)$. We use the same $k(\cdot)$ as in Example 4.1, but $H_{TD}^c = \left[0.1 \cdot T_c^{-\frac{4}{21}}, 3 \cdot T_c^{-\frac{2}{21}}\right]$ and the weight function $w(x) = I_{[2,4]}(x)$.

The semiparametric $\mathrm{MCCV}(T_v)$ selection procedure then suggests using a partially linear prediction model of the form

$$Y_{t+5} = \widehat{\beta}_1 Y_{t+4} + \widehat{\beta}_2 Y_{t+3} + \widehat{\phi}(X_{t+4}, X_{t+3}, X_{t+1}), \quad t = 1, \cdots, 288, \quad (4.25)$$

where $\widehat{\beta}_1 = 0.4129$, $\widehat{\beta}_2 = -0.3021$ and $\widehat{\phi}(\cdot)$ is a nonparametric estimator. The corresponding bandwidth chosen by (4.16) is $\widehat{h}_1 = 0.29305$.

We also consider using the nonparametric CV1 function for the same data for the case where Y_{t+i} for $i = 1, \cdots, 4$ and X_{t+j} for $j = 0, \cdots, 4$ are candidate of nonparametric regressors. The nonparametric CV1 selection function suggests the following nonparametric prediction model

$$Y_{t+5} = \widehat{m}(Y_{t+4}, Y_{t+1}, X_{t+4}, X_{t+2}, X_{t+1}), \quad t = 1, \cdots, 288, \qquad (4.26)$$

where $\widehat{m}(\cdot)$ is the usual nonparametric regression estimator as defined before. The corresponding bandwidth chosen by (4.16) is $\widehat{h}_2 = 0.14985$.

When we assume that the dependence of Y_{t+5} on Y_{t+i} for $i = 1, \cdots, 4$ and X_{t+j} for $j = 0, \cdots, 4$ is linear, the corresponding parametric $\mathrm{MCCV}(T_v)$ function suggests a linear prediction model of form

$$Y_{t+5} = \widehat{\alpha}_1 Y_{t+4} + \widehat{\alpha}_2 Y_{t+3} + \widehat{\alpha}_3 Y_{t+1} + \widehat{\alpha}_4 X_{t+3} + \widehat{\alpha}_5 X_{t+1}, \quad t = 1, \cdots, 288,$$
$$(4.27)$$

where $\widehat{\alpha}_1 = 0.3371$, $\widehat{\alpha}_2 = -0.2981$, $\widehat{\alpha}_3 = 0.0467$, $\widehat{\alpha}_4 = -0.2072$, and $\widehat{\alpha}_5 = 0.1061$.

For the whole data set, the estimated error variances for the partially linear model (4.25), the nonparametric model (4.26) and the linear model (4.27) were 0.01056, 0.02854 and 0.04389, respectively. The complete calculation for Example 4.3 took about one hour of CPU time on a PowerBook G4 System.

Example 4.3 shows that if a partially linear model among the possible partially linear models is appropriate, then the semiparametric $\mathrm{MCCV}(T_v)$ selection procedure is capable of finding it. When using both the nonparametric CV1 selection criterion and parametric $\mathrm{MCCV}(T_v)$ selection criterion, we obtain two different models for the same data set. However, the estimated error variance for the partially linear model is the smallest among those for models (4.25), (4.26) and (4.27). The findings in Example 4.3 are consistent with existing studies in that the relationship between the catch and the effort appears to be nonlinear while the current catch depends linearly on the past catch.

We acknowledge the computing expenses of the CV-based selection procedure. In our detailed simulation and computing for Examples 4.2 and 4.3, we have used some optimization algorithms, such as vectorised algorithms in the calculation of both the semiparametric $\mathrm{MCCV}(T_v)$ and the nonparametric CV1 functions of many possible candidates. The final computing time for each example is not unreasonable, but further discussion of computing algorithms is beyond the scope of this section.

In the following we illustrate Theorem 4.2 using two simulated examples. Our simulation results support the asymptotic theory and the use of the semiparametric penalty function method for partially linear model selection.

Example 4.4: Consider a nonlinear time series model of the form

$$Y_t = 0.47 U_{t-1} - 0.45 U_{t-2} + \frac{0.5 V_{t-1}}{1 + V_{t-1}^2} + e_t, \qquad (4.28)$$

where $U_t = 0.55 U_{t-1} - 0.12 U_{t-2} + \delta_t$ and $X_t = 0.3\sin(2\pi X_{t-1}) + \epsilon_t$, in which δ_t, ϵ_t and e_t are mutually independent and identically distributed with uniform distributions on $(-1, 1)$, $(-0.5, 0.5)$ and the standard

Normal distribution $N(0,1)$, respectively, U_1, U_2, V_1, V_2 are independent and identically distributed with uniform distribution on $(-1, 1)$, U_s and V_t are mutually independent for all $s, t \geq 3$, and the process $\{(\delta_t, \epsilon_t, e_t)\}$ is independent of both (U_1, U_2) and (V_1, V_2).

As model (4.28) is a special case of model (4.22) in Example 4.2, Assumptions 4.1–4.3 are satisfied immediately. For an application of Theorem 4.2, let

$$\beta = (\beta_1, \beta_2)^\tau = (0.47, -0.45)^\tau \quad \text{and} \quad g(V_{t-1}) = \frac{0.5V_{t-1}}{1 + V_{t-1}^2}.$$

To satisfy Assumption 4.4, we use $\Lambda_T(A, D) = (|A| + |D|) \cdot T^{0.5}$ as the penalty function. In addition to the choice of $\Lambda_T(A, D)$, we also considered choosing several other forms for $\Lambda_T(A, D)$. As the resulting simulated frequencies are very similar, we focus only on this choice throughout the rest of this section.

The choice of both the bandwidth interval H_T as well as the kernel function is the same as in Example 4.2. Meanwhile, Assumptions 4.5–4.8 hold trivially. For the four sample sizes $T = 52, 127, 272$ and 552, we calculated the relative frequencies of the selected parametric and nonparametric regressors in 1000 replications. Table 4.3 below reports the results of the simulation.

Table 4.3. Frequencies of semiparametric penalty model selection

Parametric & Nonparametric Set	Frequencies			
	$T = 52$	$T = 127$	$T = 272$	$T = 552$
$\{U_{t-2}, U_{t-1}, X_{t-1}\}$	0.609	0.746	0.873	0.971
$\{U_{t-2}, U_{t-1}, X_t\}$	0.350	0.235	0.123	0.028
$\{U_{t-2}, X_{t-1}, X_t\}$	0.024	0.014	0.002	0.001
$\{U_{t-1}, X_{t-1}, X_t\}$	0.017	0.005	0.002	0.000

Table 4.3 shows that the true set of regressors $\{U_{t-2}, U_{t-1}, V_{t-1}\}$ is selected with increasing frequencies from 0.554 to 0.970 as the sample size increases from $T = 52$ to $T = 552$. As expected, the other models are selected less and less frequently as the sample size increases even model $\{U_{t-2}, U_{t-1}, V_t\}$, which is one of the closest to the true model, is selected with decreasing frequencies from 0.390 to 0.030. This lends support to the efficacy of combining the penalty function method with the leave–one–out cross–validation (CV1).

Table 4.3 also shows that the proposed semiparametric model selection works well numerically when the true model is a partially linear model. As there are many existing model selection methods in the literature, in order to demonstrate the necessity of establishing a new model selection, we show in Tables 4.4 and 4.5 below that the proposed model selection method is much more effective than existing penalty function model selection methods through comparing the new method dedicated to partially linear models with the penalty function method for linear models proposed by Zheng and Loh (1997) and the conventional nonparametric leave–one–out cross–validation function CV1.

Table 4.4. Frequencies of parametric penalty model selection

Parametric Subset	Frequencies			
	$T = 52$	$T = 127$	$T = 272$	$T = 552$
$\{U_{t-2}, U_{t-1}, X_{t-1}\}$	0.055	0.142	0.383	0.684
$\{U_{t-2}, U_{t-1}\}$	0.238	0.464	0.525	0.310
$\{U_{t-1}, U_{t-2}, X_t\}$	0.024	0.013	0.009	0.003
$\{U_{t-2}\}$	0.180	0.096	0.015	0.000
$\{U_{t-1}\}$	0.193	0.112	0.020	0.000
$\{X_{t-1}\}$	0.178	0.101	0.025	0.000
$\{X_t\}$	0.067	0.024	0.001	0.000
$\{U_{t-2}, X_{t-1}\}$	0.019	0.014	0.007	0.000
$\{U_{t-1}, X_{t-1}\}$	0.022	0.024	0.008	0.000
$\{U_{t-2}, U_{t-1}, X_{t-1}, X_t\}$	0.003	0.004	0.005	0.003

Table 4.5. Frequencies of nonparametric model selection

Nonparametric Subset	Frequencies			
	$T = 52$	$T = 127$	$T = 272$	$T = 552$
$\{U_{t-2}, U_{t-1}, V_{t-1}\}$	0.103	0.288	0.464	0.652
$\{U_{t-2}, U_{t-1}, V_{t-1}, V_t\}$	0.050	0.103	0.184	0.196
$\{U_{t-2}, U_{t-1}\}$	0.135	0.205	0.194	0.117
$\{U_{t-2}\}$	0.064	0.022	0.002	0.000
$\{U_{t-1}\}$	0.102	0.035	0.004	0.000
$\{V_{t-1}\}$	0.102	0.048	0.008	0.000
$\{V_t\}$	0.059	0.011	0.000	0.000
$\{U_{t-2}, V_{t-1}\}$	0.053	0.028	0.009	0.000
$\{U_{t-1}, V_{t-1}\}$	0.058	0.060	0.014	0.002
$\{U_{t-1}, U_{t-2}, V_t\}$	0.038	0.019	0.092	0.001

The candidate variables are still $\{U_{t-2}, U_{t-1}, V_{t-1}, V_t\}$. Therefore there are $2^4 - 1 = 15$ models that can be selected. Both the penalty function method for linear model selection and the CV1 selection procedure consider all possible 15 models. As many other insignificant regressors have just tiny probabilities of being selected, Tables 4.4 and 4.5 provide only the relevant frequencies for the significant regressors.

Tables 4.4 and 4.5 imply that both the penalty function method for linear models and the conventional nonparametric CV1 method are not very effective for semiparametric models. The highest frequencies that the true model has been selected are 0.383 and 0.464, respectively when the sample size is $T \leq 272$. Although their performance improves when the sample size increases to 552, there is still a huge difference between their performance and the performance of the method proposed here.

Both the theory and Tables 4.4 and 4.5 allow us to draw the intuitively obvious conclusion that as far as the selection of a partially linear model is concerned, methods designed for purely parametric models or fully nonparametric models are not as effective as a method dedicated to partially linear models. This further emphasizes the necessity of proposing the new efficient selection method to solve problems that cannot be solved using existing selection methods for either completely linear models or fully nonparametric models.

The above simulations are based on the assumption that the true model is a partially linear time series model, for which our method is designed. But if the true model is either a purely parametric time series model or a fully nonparametric time series model, our method also performs reasonably well. Example 4.5 below considers the case where the true model is a purely parametric linear model and then applies both the parametric selection proposed by Zheng and Loh (1995) and our own semiparametric selection procedure to the model. When using the proposed semiparametric selection method, our preliminary computation suggests involving the same kernel function and bandwidth interval as used in Example 4.4 for the simulation in Example 4.5 below.

Example 4.5: Consider a linear time series model of the form

$$Y_t = 0.47U_{t-1} - 0.45U_{t-2} + 0.5V_{t-1} + e_t, \tag{4.29}$$

where

$$U_t = 0.55U_{t-1} - 0.12U_{t-2} + \delta_t \quad \text{and} \quad V_t = 0.3\sin(2\pi V_{t-1}) + \epsilon_t,$$

in which δ_t, ϵ_t and e_t are mutually independent and identically distributed with uniform distributions on $(-1, 1)$, $(-0.5, 0.5)$ and the standard Normal distribution $N(0, 1)$, respectively, U_1, U_2, V_1, V_2 are independent and identically distributed with uniform distribution on $(-1, 1)$, U_s and V_t are mutually independent for all $s, t \geq 3$, and the process $\{(\delta_t, \epsilon_t, e_t)\}$ is independent of both (U_1, U_2) and (V_1, V_2).

For the four sample sizes $T = 52, 127, 272$ and 552, we chose the penalty function $\Lambda_T(A, D) = (|A| + |D|) \, T^{0.5}$ and then calculated the relative frequencies of the selected parametric and semiparametric regressors in 1000 replications as reported in Tables 4.6 and 4.7 below.

Table 4.6. Frequencies of semiparametric model selection

Parametric & Nonparametric Subset	Frequencies			
	$T = 52$	$T = 127$	$T = 272$	$T = 552$
$\{U_{t-2}, U_{t-1}, V_{t-1}\}$	0.070	0.172	0.432	0.836
$\{U_{t-2}, U_{t-1}, V_t\}$	0.038	0.040	0.028	0.009
$\{U_{t-2}, X_{t-1}, V_t\}$	0.002	0.004	0.000	0.000
$\{U_{t-1}, V_{t-1}, V_t\}$	0.001	0.000	0.000	0.000
$\{U_{t-1}, U_{t-2}, V_{t-1}, V_t\}$	0.001	0.001	0.001	0.000
$\{U_{t-2}, V_{t-1}\}$	0.280	0.293	0.208	0.051
$\{U_{t-2}, V_t\}$	0.147	0.059	0.007	0.000
$\{U_{t-1}, V_{t-1}\}$	0.310	0.369	0.308	0.104
$\{U_{t-1}, V_t\}$	0.151	0.062	0.016	0.000

Table 4.7. Frequencies of parametric model selection

Parametric Subset	Frequencies			
	$T = 52$	$T = 127$	$T = 272$	$T = 552$
$\{U_{t-2}, U_{t-1}, V_{t-1}\}$	0.086	0.325	0.742	0.942
$\{U_{t-2}, U_{t-1}\}$	0.213	0.327	0.198	0.035
$\{U_{t-1}, U_{t-2}, V_t\}$	0.032	0.014	0.006	0.003
$\{U_{t-2}\}$	0.129	0.041	0.003	0.000
$\{U_{t-1}\}$	0.159	0.072	0.004	0.000
$\{X_{t-1}\}$	0.192	0.121	0.012	0.000
$\{V_t\}$	0.073	0.006	0.001	0.000
$\{U_{t-2}, V_{t-1}\}$	0.040	0.019	0.001	0.000
$\{U_{t-1}, V_{t-1}\}$	0.028	0.038	0.011	0.000
$\{X_{t-1}, V_t\}$	0.013	0.003	0.000	0.000
$\{U_{t-2}, U_{t-1}, V_{t-1}, V_t\}$	0.011	0.018	0.022	0.020

Tables 4.4 and 4.6 show that semiparametric penalty function method performs similarly to that of the parametric penalty function method for the cases of $T = 52, T = 127$ and $T = 272$. But in the case of $T = 552$ the former for a linear model is better than the latter for a partially linear model. While Tables 4.6 and 4.7 show that the parametric linear penalty function proposed by Zheng and Loh (1995) performs better than the semiparametric penalty function method for the case where the true model is a parametric linear model, the performance of the proposed semiparametric penalty function is comparable with that of the parametric penalty function method, particularly when T is as medium as 552.

Section 4.3 has proposed a semiparametric penalty function method for the choice of optimum regressors for both the parametric and nonparametric components. Our finite–sample simulation studies have shown that the proposed semiparametric model selection method works quite well in the case where the true model is actually a partially linear model. In addition, our simulation results have also suggested that the proposed semiparametric model selection method is a better choice than either a corresponding parametric linear model selection method or a conventional nonparametric cross–validation selection procedure when there is no information about whether the true model is parametric, nonparametric or semiparametric.

4.5 Technical notes

This section lists some technical conditions required to establish and prove Theorems 4.1 and 4.2.

4.5.1 Assumptions

Assumption 4.5. (i) Assumption 2.1 holds.

(ii) For every $D \in \mathcal{D}$, K_D is a $|D|$–dimensional symmetric, Lipschitz continuous probability kernel function with $\int ||u||^2 K_D(u)du < \infty$, and K_D has an absolutely integrable Fourier transform, where $|| \cdot ||$ denotes the Euclidean norm.

Assumption 4.6. Let S_w be a compact subset of $\mathbf{R}^p$ and w be a weight function supported on S_w and $w \leq C$ for some constant C. For every $D \in \mathcal{D}$, let $\mathbf{R}_{V,D} \subseteq \mathbf{R}^{|D|} = (-\infty, \infty)^{|D|}$ be the subset such that $V_{tD} \in \mathbf{R}_{V,D}$ and S_D be the projection of S_w in $\mathbf{R}_{V,D}$ (that is, $S_D = \mathbf{R}_{V,D} \cap S_w$).

Assume that the marginal density function, $f_D(\cdot)$, of V_{tD}, and all the first two derivatives of $f_D(\cdot)$, $g_1(\cdot)$ and $g_{A,D}(\cdot)$, are continuous on $\mathbf{R}_{V,D}$, and on S_D the density function $f_D(\cdot)$ is bounded below by C_D and above by C_D^{-1} for some $C_D > 0$, where $g_1(x) = E[Y_t|V_{tD} = x]$ and $g_{A,D}(x) = E[U_{tA}|V_{tD} = x]$ for every $A \in \mathcal{A}$ and $D \in \mathcal{D}$.

Assumption 4.7. There exist absolute constants $0 < C_1 < \infty$ and $0 < C_2 < \infty$ such that for any integer $l \geq 1$

$$\sup_x \sup_{A \in \mathcal{A}, D \in \mathcal{D}} E\left\{|Y_t - E[Y_t|(U_{tA}, V_{tD})]|^l \,|V_{tD} = x\right\} \leq C_1,$$

$$\sup_x \sup_{A \in \mathcal{A}, D \in \mathcal{D}} E\left\{||U_{tA}||^l|V_{tD} = x\right\} \leq C_2.$$

Assumption 4.8. Let

$$\begin{aligned}
\eta_t(A, D) &= U_{tA} - E[U_{tA}|V_{tD}], \\
\eta(A, D) &= (\eta_1(A, D), \ldots, \eta_T(A, D))^\tau, \\
\eta_t &= U_t - E[U_t|V_t], \quad \eta = (\eta_1, \ldots, \eta_T)^\tau, \\
Q(A, D) &= \eta(A, D)(\eta(A, D)^\tau \eta(A, D))^+ \eta(A, D)^\tau,
\end{aligned}$$

and

$$P_{1T}(A, D) = \frac{1}{T}(\eta\beta)^\mathsf{T}[I_T - Q(A, D)](\eta\beta).$$

(i) Assume that for each given $A \in \mathcal{A}_1$ and $D \in \mathcal{D}$,

$$\liminf_{T \to \infty} P_{1T}(A, D) > 0 \quad \text{in probability.}$$

(ii) As $T \to \infty$, $\frac{T_v}{T} \to 1$, $T_c = T - T_v \to \infty$ and $\frac{T^2}{T_c^2 n} \to 0$.

Assumptions 4.5–4.8 are standard in this kind of problem. See (A.1) of Cheng and Tong (1993). Due to Assumption 4.6, we need not assume that the marginal density of $\{X_t\}$ has a compact support. Assumptions 4.6–4.8 are a set of extensions of some existing conditions to the α–mixing time series case. See, for example, (A)–(E) of Zhang (1991), (A2)–(A5) of Cheng and Tong (1993), and (C.2)–(C.5) of Vieu (1994). As pointed out before, when U_t and V_t are independent, Assumption 4.8(i) imposes only an asymptotic and minimal model identifiability condition on the linear component. This means that Assumption 4.8(i) is a natural extension of condition (2.5) of Shao (1993) to the semiparametric time series setting. Assumption 4.8(ii) corresponds to conditions (3.12) and (3.22) of Shao (1993) for the linear model case. In addition, Assumption 4.8(i) is also equivalent to Assumption C of Zhang (1993) for the linear model case.

4.5.2 Technical lemma

This section lists an important technical lemma. As its proof is extremely technical, we refer it to the technical report by Gao and Tong (2005). For simplicity of notation, we assume $w(V_t) \equiv 1$ throughout the proof.

Lemma 4.1. (i) *Assume that the conditions of Theorem 4.1 hold. If $A \in \mathcal{A}_1$ and $D \in \mathcal{D}$, then there exists $R_{1T} \geq 0$ such that*

$$\text{MCCV}(A, D; h) = \frac{1}{T_v n} \sum_{S \in \mathcal{R}} \sum_{t \in S} e_t^2 + P_{1T}(A, D) + N_{1T}(D, h) + R_{1T} + o_p(1),$$
(4.30)

where R_{1T} is independent of (A, D), $P_{1T}(A, D)$ is as defined in Assumption 4.8, and

$$N_{1T}(D, h) = \begin{cases} c_1(D)\frac{1}{T_c h^{|D|}} + c_2(D)h^4 + o_p\left(\frac{1}{T_c h^{|D|}}\right) + o_p(h^4) \\ \quad \text{if } D \in \mathcal{D}_1 \text{ and } h \in H_{TD}^c; \text{ and} \\ E\left\{E[Y_t|X_{tD}] - E[Y_t|X_t]\right\}^2 + o_p(1) \\ \quad \text{if } D \notin \mathcal{D}_1 \text{ and } h \in H_{TD}^c, \end{cases}$$

in which both $c_1(D)$ and $c_2(D)$ are positive constants depending on $D \in \mathcal{D}_1$.

(ii) *Assume that the conditions of Theorem 4.1 hold. If $A \in \mathcal{A}_2$ and $D \in \mathcal{D}$, then*

$$\begin{aligned} \text{MCCV}(A, D; h) &= \frac{1}{T_v n} \sum_{S \in \mathcal{R}} \sum_{t \in S} e_t^2 + \frac{d_A}{T_c}\sigma_0^2 + N_{1T}(D, h) \\ &\quad + o_p\left(\frac{1}{T_c}\right), \end{aligned}$$
(4.31)

where $N_{1T}(D, h)$ is as defined as above depending on whether $D \in \mathcal{D}_1$ or not.

The proof of Lemma 4.1 is relegated to Appendix B of Gao and Tong (2005). As pointed out earlier, $P_{1T}(A, D) = P_{1T}(A)$ depends only on A when U_t and V_t are independent. For this case, Equations (4.30) and (4.31) suggest naturally that the selection of A and D can be done independently.

4.5.3 Proofs of Theorems 4.1 and 4.2

Proof of Theorem 4.1: It follows from existing results in nonparametric regression (Vieu 1994) that for each given D, there exists $\bar{h}_D =$

$c_D T_c^{-\frac{1}{4+|D|}}$ such that

$$
N_{1T}(D) = \min_{h \in H_{TD}^c} N_{1T}(D, h) = \begin{cases} C_D T_c^{-\frac{4}{4+|D|}} + o_p\left(T_c^{-\frac{4}{4+|D|}}\right) \\ \text{if } D \in \mathcal{D}_1; \text{ and} \\ E\left\{E[Y_t|X_{tD}] - E[Y_t|X_t]\right\}^2 + o_p(1) \\ \text{if } D \notin \mathcal{D}_1, \end{cases}
$$

in which c_D and C_D are positive constants possibly depending on D.

Let $\mathrm{MCCV}(A, D) = \min_{h \in H_{TD}^c} \mathrm{MCCV}(A, D; h)$. In view of (4.30) and (4.31), it is known that for each given (A, D),

$$
\mathrm{MCCV}(A, D) = \begin{cases} \frac{1}{T_v n} \sum_{S \in \mathcal{R}} \sum_{t \in S} e_t^2 + P_{1T}(A, D) + N_{1T}(D) \\ + R_{1T} + o_p(1) \text{ if } A \in \mathcal{A}_1 \text{ and } D \in \mathcal{D}; \text{ and} \\ \frac{1}{T_v n} \sum_{S \in \mathcal{R}} \sum_{t \in S} e_t^2 + \frac{d_A}{T_c} \sigma_0^2 + N_{1T}(D) + o_p\left(\frac{1}{T_c}\right) \\ \text{if } A \in \mathcal{A}_2 \text{ and } D \in \mathcal{D}. \end{cases}
$$

This, along with $P_{1T}(A_*, D_*) = 0$, implies

$$
\mathrm{MCCV}(A, D) - \mathrm{MCCV}(A_*, D_*) = \begin{cases} P_{1T}(A, D) + N_{1T}(D) - N_{1T}(D_*) \\ + o_p(1) \text{ if } A \in \mathcal{A}_1 \text{ and } D \in \mathcal{D}; \text{ and} \\ \frac{(d_A - d_{A_*})}{T_c} \sigma_0^2 + N_{1T}(D) - N_{1T}(D_*) \\ + o_p\left(\frac{1}{T_c}\right) \text{ if } A \in \mathcal{A}_2 \text{ and } D \in \mathcal{D}. \end{cases}
$$

Using the fact that Assumption 4.1(ii) implies $|D| > |D_*|$ for $D \in \mathcal{D}_1$, we have for T large enough

$$
T_c^{\frac{4}{4+|D_*|}} (N_{1T}(D) - N_{1T}(D_*)) = C_D T_c^{\frac{4(|D|-|D_*|)}{(4+|D|)(4+|D_*|)}} - C_{D_*}
$$
$$
+ o_p\left(T_c^{\frac{4(|D|-|D_*|)}{(4+|D|)(4+|D_*|)}}\right) > 0 \quad (4.32)
$$

in probability.

On the other hand, for every $D \in \mathcal{D}_1^c = \mathcal{D} - \mathcal{D}_1$ we have for T large enough

$$
N_{1T}(D) - N_{1T}(D_*) = E\left(E[Y_t|X_{tD}] - E[Y_t|X_t]\right)^2 + o_p(1) > 0 \quad (4.33)
$$

in probability.

Using Assumption 4.8(i) for $A \in \mathcal{A}_1$ and the fact that Assumption 4.1(ii) implies $d_A > d_{A_*}$ for $A \in \mathcal{A}_2$, the proof of $\lim_{T \to \infty} P(\widehat{A} = A_*, \widehat{D} = D_*) = 1$ then follows from (4.32) and (4.33).

Let $\widehat{h} = \bar{h}_{\widehat{D}}$, $c_* = c_{D_*}$ and $h_* = \bar{h}_{D_*} = c_* T_c^{-\frac{1}{4+|D_*|}}$. Then the proof of $\frac{\widehat{h}}{h_*} \to_p 1$ follows immediately.

Proof of Theorem 4.2: Let $RSS(A, D) = \min_{h \in H_{TD}} RSS(A, D; h)$. We first write $RSS(A, D)$ as

$$RSS(A, D) = \begin{cases} \sum_{t=1}^{T} e_t^2 + T \cdot P_{1T}(A, D) + T \cdot N_{1T}(D, \bar{h}_D) \\ + R_{1T} + o_p(T) \text{ if } A \in \mathcal{A}_1, D \in \mathcal{D}; \text{ and} \\ \sum_{t=1}^{T} e_t^2 + d_A \sigma^2 + T \cdot N_{1T}(D, \bar{h}_D) + o_p(1) \\ \text{if } A \in \mathcal{A}_2, D \in \mathcal{D}. \end{cases}$$

It follows immediately from $P_{1T}(A_*, D_*) = 0$ that
$RSS(A, D) - RSS(A_*, D_*)$

$$= \begin{cases} T \cdot P_{1T}(A, D) - d_{A_*} \sigma^2 + T(N_{1T}(D, \bar{h}_D) - N_{1T}(D_*, \bar{h}_{D_*})) \\ + o_p(T) \text{ if } A \in \mathcal{A}_1, D \in \mathcal{D}; \text{ and} \\ (d_A - d_{A_*})\sigma^2 + T(N_{1T}(D, \bar{h}_D) - N_{1T}(D_*, \bar{h}_{D_*})) + o_p(1) \\ \text{if } A \in \mathcal{A}_2, D \in \mathcal{D}. \end{cases}$$

In order to complete the proof, we introduce the following symbols:

$$\begin{aligned} \overline{RSS} &= RSS(A, D) - RSS(A_*, D_*), \\ \overline{\Lambda}_T &= \Lambda_T(A, D) - \Lambda_T(A_*, D_*), \\ \overline{N}_{1T} &= N_{1T}(D, \bar{h}_D) - N_{1T}(D_*, \bar{h}_{D_*}), \\ \overline{d}(A, A_*) &= d_A - d_{A_*}. \end{aligned} \tag{4.34}$$

If $A \in \mathcal{A}_2$ and $D \in \mathcal{D}$, then we have as $T \to \infty$,

$$1 - P\left\{\overline{RSS} + \overline{\Lambda}_T \hat{\sigma}^2 > 0\right\}$$

$$= P\left\{(d_A - d_{A_*})\sigma^2 + T \overline{N}_{1T} + o_p(1) + \overline{\Lambda}_T \hat{\sigma}^2 \leq 0\right\}$$

$$= P\left\{\overline{N}_{1T} \leq -\frac{\overline{d}(A, A_*)}{T}\sigma^2 - \frac{\overline{\Lambda}_T}{T}\hat{\sigma}^2\right\}$$

$$\leq P\left\{\overline{N}_{1T} \leq -\frac{\overline{d}(A, A_*)}{T}\sigma^2 - \frac{\overline{\Lambda}_T}{T} \cdot \frac{1}{2}\sigma^2\right\} + o(1)$$

$$\leq P\left\{\overline{N}_{1T} \leq -\frac{\Lambda_T(A, D)}{T}\left(1 - \frac{\Lambda_T(A_*, D_*)}{\Lambda_T(A, D)}\right) \cdot \frac{1}{2}\sigma^2\right\} + o(1) \to 0$$

because of $\overline{N}_{1T} > 0$ for all $A \in \mathcal{A}_1$ and either $D \in \mathcal{D}_1$ or $D \in \mathcal{D} - \mathcal{D}_1$.

If $A \in \mathcal{A}_1$ and $D \in \mathcal{D}$, then we obtain as $T \to \infty$,

$$1 - P\left\{\overline{RSS} + (\Lambda_T(A, D) - \Lambda_T(A_*, D_*))\hat{\sigma}^2 > 0\right\}$$

$$= P\left\{P_{1T}(A, D) + \overline{N}_{1T} + \frac{R_{1T} - d_{A_*}\sigma^2 + \overline{\Lambda}_T \hat{\sigma}^2}{T} + o_P(1) \leq 0\right\}$$

$$= P\left\{P_{1T}(A, D) + \overline{N}_{1T} \leq -\frac{o_p(T)}{T} + \frac{d_{A_*}\sigma^2}{T} - \frac{R_{1T}}{T} - \frac{\overline{\Lambda}_T \hat{\sigma}^2}{T}\right\} + o(1)$$

$$= P\left\{ P_{1T}(A, D) + \overline{N}_{1T} \le -\frac{\Lambda_T(A, D)}{T}\left(1 - \frac{\Lambda_T(A_*, D_*)}{\Lambda_T(A, D)}\right) \cdot \frac{1}{2}\sigma^2\right\}$$
$$+ o(1) \to 0$$

because of $\liminf_{T\to\infty} P_{1T}(A, D) > 0$ and $\overline{N}_{1T} > 0$ for all $D \in \mathcal{D}_1$ or $D \in \mathcal{D} - \mathcal{D}_1$.

Consequently, we have as $T \to \infty$,

$$1 \ge P(\widetilde{A} = A_*, \widetilde{D} = D_*) \ge P\left\{\overline{RSS} + \overline{\Lambda}_T\,\widehat{\sigma}^2 > 0\right\} \to 1. \qquad (4.35)$$

This completes the proof of the first part of Theorem 4.2. The second part is the same as in the proof of Theorem 4.1.

4.6 Bibliographical notes

Since the work by Stone (1977), Zhang (1991), Bickel and Zhang (1992), Cheng and Tong (1992, 1993), Tjøstheim and Auestad (1994a, 1994b), Vieu (1994), Yao and Tong (1994), Vieu (1995), Shao (1997), Shi and Tsai (1999), Yang (1999), Tjøstheim (1999), Tschernig and Yang (2000), Härdle, Liang and Gao (2000), Vieu (2002), Avramidis (2005), and others have established various kernel–based selection criteria in nonparametric regression for both the independent and time series cases.

In addition to such kernel–based selection criteria, some other variable selection criteria have also been proposed and studied recently by Shi and Tsai (1999) through involving the spline smoothing method, Fan and Li (2001, 2002) using the so–called penalized likelihood method, and Shively and Kohn (1997), Shively, Kohn and Wood (1999), Kohn, Marron and Yau (2000), Wood *et al.* (2002), Wong, Carter and Kohn (2003), Yau and Kohn (2003), and Yau, Kohn and Wood (2003) based on the Bayesian approach.

CHAPTER 5

Continuous–Time Diffusion Models

5.1 Introduction

This chapter demonstrates how to apply the estimation and specification testing procedures discussed in Chapters 2 and 3 to certain model estimation and specification testing problems in continuous–time models. While we use two financial data sets to illustrate the applicability of these estimation and testing procedures to continuous–time models, both the theory and methodology discussed in this chapter are applicable to model some other nonfinancial data problems. The main material of this chapter is mainly based on the joint work by Casas and Gao (2005) and Arapis and Gao (2006).

5.1.1 Parametric models

The application of continuous–time mathematics to the field of finance dates back to 1900 when Louis Bachelier wrote a dissertation in Bachelier (1900) on the theory of speculation. Since Bachelier, the continuous–time approach to pricing assets such as derivative securities has evolved into a fundamental finance tool. The recent rapid expansion of asset pricing theory may be largely attributable to the seminal work of Merton (1973) and Black and Scholes (1973). Their work changed the way in which finance asset valuation was viewed by practitioners, consequently laying the foun-dation for the theory of pricing derivative securities. Many papers have since been written on the valuation of derivatives, creating important extensions to the original model.

A time series model used extensively in finance is the continuous-time diffusion, or Itô process. In modeling the dynamics of the short-term riskless rate process $\{r_t\}$, for example, the applicable diffusion process is

$$dr_t = \mu(r_t)dt + \sigma(r_t)dB_t, \qquad (5.1)$$

where $\mu(\cdot) = \mu(\cdot, \theta)$ and $\sigma(\cdot) = \sigma(\cdot, \theta)$ are the drift and volatility functions of the process, respectively, and can be indexed by θ, a vector of unknown parameters, and B_t is the standard Brownian motion. The diffusion function is also referred to as the instantaneous variance. The model developed by Merton specified the drift and diffusion functions as constant. This assumption has since been relaxed by most researchers interested in refining the model in order to describe the behavior of interest rates. The prices generated by such modified models are generally believed to better reflect those observed in the market.

A vast array of models has been studied in the literature. The simplest model $dr = \alpha(\beta - r)dt + \sigma dB$ proposed by Vasicek (1977) was used to derive a discount bond price model. Unlike the model developed by Merton (1973) and Black and Scholes (1973), whose respective process follow Brownian motion with drift and Geometric Brownian Motion (GBM), Vasicek (1977) utilized the Ornstein-Uhlenbeck process. This model has the feature of mean-reversion, where the process tends to be pulled to its long–run trend of β with force α. This force is proportional to the deviation of the interest rate from its mean. This model specifies the volatility of the interest rate as being constant. By definition, the volatility function generates the erratic fluctuations of the process around its trend. Cox, Ingersoll and Ross (CIR) (1985) proposed using model $dr = \alpha(\beta - r)dt + \sigma r^{1/2}dB$ to model term–structure. It is the square root process. Not only does the drift have mean–reversion, but the model also implies the volatility $\sigma(\cdot)$, of the process increases at a rate proportional to $\sqrt{r}$. Thus the diffusion increases at a rate proportional to r. Model $dr = \alpha(\beta - r)dt + \sigma r dB$ (see Brennan and Schwartz 1980) was developed to price convertible bonds. It not only possesses the mean–reversion property, but the model also implies that the instantaneous variance $\sigma^2(\cdot)$ of the process increases at a rate proportional to r^2. Model $dr = r\{\kappa - (\sigma^2 - \kappa\alpha)r\}dt + \sigma r^{3/2}dB$ is the inverse of the CIR process discussed in Ahn and Gao (1999) and Aït-Sahalia (1999). Model $dr = \kappa(\alpha - r)dt + \sigma r^\rho dB$ is the constant elasticity of volatility model proposed in Chan et al. (1992). The nonlinear drift model $dr = (\alpha_{-1}r^{-1} + \alpha_0 + \alpha_1 r + \alpha_2 r^2)dt + \sigma r^{3/2}dB$ was proposed in Aït-Sahalia (1996a).

As well as the recent developments made in the application of continuous-time diffusion processes to the finance world, there has also been much work done in the adoption of statistical methods for the estimation of these continuous-time models. The main estimation techniques encountered in the majority of the literature (see Sundaresan 2001) include maximum likelihood (ML), generalized method of moments (GMM) and, more recently, nonparametric approaches. ML and GMM both require us to firstly parameterize the underlying model of interest. That is, we apply these methods to estimate the parameters of the diffusion process, such that they are consistent with the restrictions we have imposed on the model by the parameterizations. This is comparable to fitting a linear regression to nonlinear phenomena for reasons of convenience. It thus seems reasonable that we look for an approach that places the fewest restrictions on models so that we have empirical rather than analytical tractability.

5.1.2 Nonparametric models

Empirical researchers have recently shown that nonparametric methods may be good alternatives to parametric methods in various cases. Its only prerequisite is that accurate data are used. Such an approach is useful when approximating very general distributions, and has the additional advantage of not requiring the functional form specification of the drift and diffusion functions in our model of the short-term riskless rate (5.1). By leaving the diffusion process unspecified, the resulting functional forms specified by this method should result in a process that follows asset price data closely. This method requires a smooth density estimator of the marginal distribution $\pi(\cdot)$ and utilization of a property of (5.1), similar to that of a normal random variable whose distribution is explained entirely by its first two moments, to characterize the marginal and conditional densities of the interest rate process. The first two moments of the normal distribution are its mean and variance. For the case of the diffusion process, they are the drift and diffusion functions. Thus, the functions are formed such that they are consistent with the observed distribution of the available data.

Aït-Sahalia (1996a) was among the first to pioneer the nonparametric approach. The paper noted, as with Chan et al. (1992) and Ahn and Gao (1999), that one of the most important features of the process given by (5.1) in its ability to accurately model the term structure of interest rates is the specification of the diffusion function $\sigma^2(\cdot)$. By qualifying the restriction on the drift function $\mu(\cdot)$, to the linear parametric class $\mu(r_t; \theta) = \beta(\alpha - r_t)$, which is consistent with the majority of prior research, the form of the diffusion function is left unspecified and estimated nonparametrically. Jiang and Knight (1997), however, argued that this is effectively a semiparametric approach because of the linear restriction imposed on the drift function. Jiang and Knight (1997) were able to develop an identification and estimation procedure for both the drift and diffusion functions of a general Itô diffusion process. They too have used nonparametric kernel estimators of the marginal density function based on discretely sampled data and the property of the Itô diffusion process, analogous to that used by Aït-Sahalia (1996a). In contrast to Aït-Sahalia

(1996a), the drift function is left unspecified. Jiang and Knight (1997) suggested that the diffusion term can be identified first because it is of lower order than the drift. It is noted that the diffusion term is of order $\sqrt{dt}$ whereas the drift term is of order dt. These estimators as with that of Aït-Sahalia (1996a) are shown to be pointwise consistent and asymptotically normal.

Like most existing studies, we apply the Euler first–order scheme to approximate model (5.1) by a discretized alternative of the form

$$r_{(t+1)\Delta} - r_{t\Delta} = \mu(r_{t\Delta})\Delta + \sigma(r_{t\Delta}) \cdot (B_{(t+1)\Delta} - B_{t\Delta}), \ t = 1, 2, \cdots, T, \ (5.2)$$

where Δ is the time between successive observations and T is the size of observations. In theory, we study asymptotic properties of our nonparametric estimators for the case where Δ is either varied according to T or small but fixed. In applications, we look at the case where Δ is small but fixed, since most continuous–time models in finance are estimated with monthly, weekly, daily, or higher frequency observations. For the case where Δ is varied according to T, we establish some novel asymptotic properties for both the drift and diffusion estimators. Similarly to existing studies (Bandi and Phillips 2003; Nicolau 2003), we show that nonparametric estimators may be inconsistent when Δ is chosen as fixed. It should be noted that using a higher–order approximate version rather than (5.2) may not be optimal in terms of balancing biases between drift and diffusion estimation as studied in Fan and Zhang (2003). We therefore use the first–order approximate model (5.2) to study our nonparametric estimation throughout this chapter.

5.1.3 Semiparametric models

Unlike the work by Aït-Sahalia (1996a) and Jiang and Knight (1997), in order to avoid undersmoothing, Arapis and Gao (2006) have proposed an improved and simplified nonparametric approach to the estimation of both the drift and diffusion functions and established the mean integrated square error (MISE) of each nonparametric estimator for the case where Δ is either varied according to T or small but fixed. The authors have then applied the proposed nonparametric approach to (i) the Federal Funds rate data, sampled monthly between January 1963 and December 1998 and (ii) the Eurodollar deposit rates, bid-ask midpoint and sampled daily from June 1, 1973 to February 25, 1995. Three nonparametric and semiparametric methods for estimating the drift and diffusion functions are established. For each data set, these estimators have been computed. Bandwidth selection is both difficult and critical to the application of the nonparametric approach. After empirical comparisons the authors have suggested for each given set of data, the best fitting model and bandwidth which produces the most acceptable results. The authors' study shows that the imposition of the parametric linear mean–reverting drift does in fact affect the estimation of the diffusion function. Differences between the three diffusion estimators suggest the drift function may have a greater effect on pricing derivatives than what is quoted in the literature.

Furthermore, this chapter considers the following two semiparametric models:

$$r_{(t+1)\Delta} - r_{t\Delta} \;=\; \mu(r_{t\Delta}, \theta)\Delta + \sigma(r_{t\Delta}) \cdot (B_{(t+1)\Delta} - B_{t\Delta}), \quad (5.3)$$
$$r_{(t+1)\Delta} - r_{t\Delta} \;=\; \mu(r_{t\Delta})\Delta + \sigma(r_{t\Delta}, \vartheta) \cdot (B_{(t+1)\Delta} - B_{t\Delta}), \quad (5.4)$$

where θ and ϑ are vectors of unknown parameters. Estimation problems for models (5.3) and (5.4) have been studied in Kristensen (2004). This chapter thus focuses on semiparametric tests for parametric specification of the diffusion function in model (5.3) and the drift function of model (5.4). Two of the most relevant papers to the testing part of the current chapter are Casas and Gao (2005) and Arapis and Gao (2006).

Other closely related papers include Chen and Gao (2005) and Hong and Li (2005), who both developed nonparametric specification tests for transitional densities of continuous–time diffusion models. These two recent studies are motivated by the fact that unlike the marginal density, the transitional density can capture the full dynamics of a diffusion process when interest is on the specification of a diffusion process. Since such transitional density specification may not directly imply the specification of the drift function, the test proposed in Arapis and Gao (2006) on the specification of the parametric drift function is more direct to answer the question of whether there is any nonlinearity in the drift. The paper by Arapis and Gao (2006) has the following features. First, it establishes that the size of the test is asymptotically correct under any model in the null. Second, it shows that the test is asymptotically consistent when the null is false. Third, the implementation of the proposed test uses a range of bandwidth values instead of using an estimation optimal value based on a cross–validation selection criterion. Fourth, the employment of the proposed test is based on a simulated p–value rather than an asymptotic critical value of the standard normality. It should also be pointed out that the proposed test is applicable to either the case where Δ is either varied according to T or the case where Δ is small but fixed.

5.2 Nonparametric and semiparametric estimation

The nonparametric approach to density estimation allows modeling of data where no priori ideas about the data exist. Given T discrete interest rate observations with sampling interval (equivalently the time between successive observations) Δ, the kernel density estimate of the marginal density is given by

$$\widehat{\pi}(r) = \frac{1}{T} \sum_{t=1}^{T} \frac{1}{h} K\left(\frac{r - r_{t\Delta}}{h}\right), \tag{5.5}$$

where $K(\cdot)$ is the kernel function and h is the kernel bandwidth. By comparing the nonparametric marginal density, drift and diffusion estimates acquired by use of some existing "good" bandwidth values, we can suggest the most appropriate bandwidth to use for each of our two different sets of financial data. Whereas bandwidth selection is critical for optimal results, the selection of the kernel does not have a significant bearing on the overall result (see §3.2.6 of Fan and Gijbels 1996). We therefore utilize the normal kernel function $K(x) = \frac{1}{\sqrt{2\pi}} \exp\{-\frac{x^2}{2}\}$ throughout this chapter. Our nonparametric marginal density estimate is

$$\widehat{\pi}(r) = \frac{1}{\sqrt{2\pi}hT} \sum_{t=1}^{T} \exp\left\{\frac{-(r - r_{t\Delta})^2}{2h^2}\right\}. \tag{5.6}$$

Asymptotic consistency results about $\widehat{\pi}(\cdot)$ may be found from Boente and Fraiman (1988). From the Fokker–Planck equation (see (2.2) of Aït-Sahalia 1996a), we can obtain

$$\frac{d^2}{dr^2}(\sigma^2(r)\pi(r)) = 2\frac{d}{dr}(\mu(r)\pi(r)). \tag{5.7}$$

Integrating and rearranging (5.7) yields

$$\mu(r) = \frac{1}{2\pi(r)} \frac{d}{dr}\left[\sigma^2(r)\pi(r)\right]. \tag{5.8}$$

Or, alternatively, integrating (5.7) twice yields

$$\sigma^2(r) = \frac{2}{\pi(r)} \int_0^r \mu(x)\pi(x)dx \tag{5.9}$$

using the condition that $\pi(0) = 0$. These equations allow us to estimate the drift, $\mu(\cdot)$ given a specification of the diffusion, $\sigma^2(\cdot)$ and marginal density, $\pi(\cdot)$, or the diffusion term given the drift and marginal density estimates.

We now start discussing three estimation methods. The first one is a kind of semiparametric estimation as proposed in Aït-Sahalia (1996a). The estimators of the next two methods rely on various alternative interpretations of the diffusion process. Neither of them place any restrictions on the drift nor the diffusion functions.

5.2.1 Method 1

For this method, we construct the diffusion function after placing the common mean–reverting parameterization $\mu(r;\theta) = \beta(\alpha - r)$ on the drift, similarly to the approach taken by Aït-Sahalia (1996a). This restriction will allow us to see how the diffusion function is affected when compared with purely nonparametric estimators to be discussed in Methods 2 and 3 below.

The parameters β and α are estimated by the ordinary least squares (OLS) method and denoted by $\widehat{\beta}$ and $\widehat{\alpha}$, respectively. This suggests estimating $\mu(r;\theta)$ by

$$\widehat{\mu}_1(r) = \mu(r;\widehat{\theta}) = \widehat{\beta}(\widehat{\alpha} - r). \tag{5.10}$$

The estimated mean–reverting drift term can now be substituted into (5.9) together with the normal kernel density estimator for $\pi(\cdot)$ to construct the first estimator, $\widehat{\sigma}_1^2(\cdot)$, of $\sigma^2(\cdot)$ below. More complete mathematical details of this derivation are relegated to Section 5.5.

From (5.9) and (5.10), we define a semiparametric estimator of $\sigma^2(r)$ of the form

$$
\begin{aligned}
\widehat{\sigma}_1^2(r) &= \frac{2}{\widehat{\pi}(r)} \int_0^r \widehat{\mu}_3(u)\widehat{\pi}(u)du = \frac{2}{\widehat{\pi}(r)} \int_0^r \mu(u;\widehat{\theta})\widehat{\pi}(u)du \\
&= \frac{2}{T\widehat{\pi}(r)} \left\{ \widehat{\beta} \sum_{t=1}^T (\widehat{\alpha} - r_{t\Delta}) \left[\Phi\left(\frac{r - r_{t\Delta}}{h}\right) - \Phi\left(\frac{-r_{t\Delta}}{h}\right) \right] \right. \\
&\quad + \left. \frac{h\widehat{\beta}}{\sqrt{2\pi}} \sum_{t=1}^T \left[\exp\left(-\frac{(r - r_{t\Delta})^2}{2h^2}\right) - \exp\left(-\frac{r_{t\Delta}^2}{2h^2}\right) \right] \right\}, \tag{5.11}
\end{aligned}
$$

where $\Phi(\cdot)$ is the cumulative distribution function of the standard Normal random variable.

As can be seen, $\widehat{\sigma}_1^2(r)$ has an explicit and computationally straightforward expression due to the use of the standard Normal kernel function. Let $\widehat{V}_1(r) = \widehat{\sigma}_1^2(r)\widehat{\pi}(r)$. For the nonparametric estimator $\widehat{V}_1(r)$, we establish the following theorem. Its proof is relegated to Section 5.5.

Theorem 5.1. *Assume that Assumptions 5.1–5.3 listed in Section 5.5* *below hold.* Then

$$E\left\{\int\left[\widehat{V}_1(r)-V(r)\right]^2 dr\right\}=4h^4\int\left(\int_0^r\beta(\alpha-u)\pi''(u)du\right)^2 dr$$

$$+\frac{4\beta^2}{Th}\int\left(\int_0^r\int_0^r(\alpha-u)(\alpha-v)L\left(\frac{u-v}{h}\right)\pi(v)dvdu\right)dr.$$

When $h=c\,T^{-\frac{1}{5}}$ *for some* $c>0$, *we have*

$$E\left\{\int\left[\widehat{V}_1(r)-V(r)\right]^2 dr\right\}=C_1T^{-\frac{4}{5}}+o\left(T^{-\frac{4}{5}}\right),\tag{5.12}$$

where $C_1>0$ *is a constant.*

As expected, asymptotic properties of this kind of semiparametric estimator $\widehat{V}_1(r)$ do not depend on Δ. This is mainly because we need only to use discrete data rather than the discretized version (5.2). When both the forms of the drift and diffusion functions are unknown nonparametrically, we need to use the discretized version (5.2) to approximate model (5.1). The following two methods rely on relationships between the drift, diffusion and marginal density functions which alternatively describe the usual diffusion process.

5.2.2 Method 2

The drift and diffusion functions can be alternatively interpreted as

$$\mu(r_t)\quad=\quad\lim_{\delta\to0}E\left[\frac{r_{t+\delta}-r_t}{\delta}|r_t\right]\quad\text{and}\tag{5.13}$$

$$\sigma^2(r_t)\quad=\quad\lim_{\delta\to0}E\left[\frac{[r_{t+\delta}-r_t]^2}{\delta}|r_t\right]\tag{5.14}$$

for all $0<t<\infty$.

Stanton (1997) refered to these right–hand conditional expectations of (5.13) and (5.14) as the first order approximations to $\mu(\cdot)$ and $\sigma^2(\cdot)$. The author constructed a family of approximations to the drift and diffusion functions and estimates the approximations nonparametrically. Equations (5.13) and (5.14) support the use of the approximate alternative (5.2) to model (5.1) when Δ is small.

Now, (5.13) suggests estimating $\mu(\cdot)$ by

$$\widehat{\mu}_2(r)=\frac{\sum_{t=1}^{T-1}K\left(\frac{r-r_{t\Delta}}{h}\right)\left(\frac{r_{(t+1)\Delta}-r_{t\Delta}}{\Delta}\right)}{\sum_{t=1}^{T}K\left(\frac{r-r_{t\Delta}}{h}\right)}.\tag{5.15}$$

Multiplying the numerator and denominator by $\frac{1}{Th}$ gives

$$\widehat{\mu}_2(r) = \frac{\frac{1}{\Delta Th}\sum_{t=1}^{T-1}K\left(\frac{r-r_{t\Delta}}{h}\right)\left(r_{(t+1)\Delta}-r_{t\Delta}\right)}{\frac{1}{Th}\sum_{t=1}^{T}K\left(\frac{r-r_{t\Delta}}{h}\right)} \tag{5.16}$$

$$= \frac{1}{\Delta Th\widehat{\pi}(r)\sqrt{2\pi}}\sum_{t=1}^{T-1}\exp\left(-\frac{(r-r_{t\Delta})^2}{2h^2}\right)\cdot\left(r_{(t+1)\Delta}-r_{t\Delta}\right),$$

when $K(x) = \frac{1}{\sqrt{2\pi}}\exp\{-\frac{x^2}{2}\}$.

Similarly, by (5.14) we estimate $\sigma^2(\cdot)$ by

$$\widehat{\sigma}_2^2(r) = \frac{1}{\Delta Th\,\widehat{\pi}(r)\sqrt{2\pi}}\sum_{t=1}^{T-1}\exp\left(-\frac{(r-r_{t\Delta})^2}{2h^2}\right)\cdot\left(r_{(t+1)\Delta}-r_{t\Delta}\right)^2. \tag{5.17}$$

Let $\widehat{m}_2(r) = \widehat{\mu}_2(r)\widehat{\pi}(r)$, $m(r) = \mu(r)\pi(r)$, $\widehat{V}_2(r) = \widehat{\sigma}_2^2(r)\widehat{\pi}(r)$, and $V(r) = \sigma^2(r)\pi(r)$. Since $\frac{\widehat{m_2(r)}}{\pi(r)}$ and $\widehat{\mu}_2(r)$ have the same asymptotic property for the MISE, we only establish the MISE for $\widehat{m}_2(r)$ below. The same reason applies to explain why $\widehat{V}_2(r)$ has been introduced. We now have the following propositions, and their proofs are relegated to Section 5.5.

Theorem 5.2. *Assume that Assumptions 5.1–5.3 listed in Section 5.5 below hold. Then*

$$E\left\{\int[\widehat{m}_2(r)-m(r)]^2\,dr\right\} = \frac{h^4}{4}\int(m''(r))^2\,dr$$

$$+\frac{1}{Th}\cdot\frac{1}{2\sqrt{\pi}}\int\left(\mu^2(r)+\sigma^2(r)\Delta^{-1}\right)\pi(r)dr$$

$$+o\left(\frac{\Delta^2}{Th}\right)+o(h^4).$$

(i) *When Δ is fixed but $h = c\,T^{-\frac{1}{5}}$ for some $c > 0$, we have*

$$E\left\{\int[\widehat{m}_2(r)-m(r)]^2\,dr\right\} = C_2\cdot T^{-\frac{4}{5}}+o\left(T^{-\frac{4}{5}}\right), \tag{5.18}$$

where $C_2 > 0$ is a constant.

(ii) *When $\Delta = c_1\,h^2$ and $h = c_2\,T^{-\frac{1}{7}}$ for some $c_1 > 0$ and $c_2 > 0$, we have*

$$E\left\{\int[\widehat{m}_2(r)-m(r)]^2\,dr\right\} = C_3\cdot T^{-\frac{4}{7}}+o\left(T^{-\frac{4}{7}}\right), \tag{5.19}$$

where $C_3 > 0$ is a constant.

Theorem 5.3. *Assume that Assumptions 5.1–5.3 listed in Section 5.5 below hold. Then*

$$E\left\{\int \left[\widehat{V}_2(r) - V(r)\right]^2 dr\right\} = o\left(\frac{1}{Th}\right) + o(h^4)$$

$$+\frac{1}{Th} \cdot \frac{1}{2\sqrt{\pi}} \int \left(\mu^4(r) + \frac{3}{\Delta^2}\sigma^4(r) + \frac{6}{\Delta}\mu^2(r)\sigma^2(r)\right) \pi(r)dr$$

$$+\int \left(\Delta\mu^2(r)\pi(r) + \frac{\Delta^2}{2}h^2 \left(\mu^2(r)\pi(r)\right)'' + \frac{h^2}{2}V''(r)\right)^2 dr.$$

(i) *When Δ is fixed but $h = d\, T^{-\frac{1}{5}}$ for some $d > 0$, we have*

$$E\left\{\int \left[\widehat{V}_2(r) - V(r)\right]^2 dr\right\} = C_4 T^{-\frac{4}{5}} + C_5\Delta^2 + o\left(T^{-\frac{4}{5}}\right), \qquad (5.20)$$

where $C_4 > 0$ and $C_5 > 0$ are constants.

(ii) *When $\Delta = d_1\, h^2$ and $h = d_2\, T^{-\frac{1}{7}}$ for some $d_1 > 0$ and $d_2 > 0$, we have*

$$E\left\{\int \left[\widehat{V}_2(r) - V(r)\right]^2 dr\right\} = C_6 T^{-\frac{4}{7}} + o\left(T^{-\frac{4}{7}}\right), \qquad (5.21)$$

where $C_6 > 0$ is a constant.

Theorems 5.2 and 5.3 show that while $\widehat{m}_2(r)$ attains the optimal MISE rate of $T^{-\frac{4}{5}}$, $\widehat{V}_2(r)$ is not even consistent when Δ is small but fixed. When the drift of model (5.1) vanishes (i.e., $\mu(r) \equiv 0$), however, the optimal MISE rate of $T^{-\frac{4}{5}}$ can also be achieved for $\widehat{V}_2(r)$.

By making use of Equations (5.8) and (5.9) with the pair $(\widehat{\mu}_2(r), \widehat{\sigma}_2^2(r))$, we can forego the prior necessities of having to specify either of the otherwise unknown functions, $\mu(\cdot)$ and $\sigma^2(\cdot)$, in order to calculate the other.

5.2.3 Method 3

This method adopts a similar approach to that taken by Jiang and Knight (1997). They have estimated $\sigma^2(\cdot)$ by

$$\widehat{\sigma}_{JK}^2(r) = \frac{\sum_{t=1}^{T-1} TK\left(\frac{r_{t\Delta_T} - r}{h}\right) \left(r_{(t+1)\Delta_T} - r_{t\Delta_T}\right)^2}{\sum_{t=1}^{T} NK\left(\frac{r_{t\Delta_T} - r}{h}\right)}, \qquad (5.22)$$

which is comparable with $\hat{\sigma}_2^2(\cdot)$, where N is the time length, Δ_T depends on T and $\Delta_T \to 0$ as $T \to \infty$. Jiang and Knight (1997) estimated the drift by

$$\hat{\mu}_{JK}(r) = \frac{1}{2}\left(\frac{d\hat{\sigma}_{JK}^2(r)}{dr} + \hat{\sigma}_{JK}^2(r)\frac{\sum_{t=1}^{T}\frac{1}{h}K'\left(\frac{r_{t\Delta_T}-r}{h}\right)}{\sum_{t=1}^{T}K\left(\frac{r_{t\Delta_T}-r}{h}\right)}\right), \qquad (5.23)$$

but as we shall see, this is unnecessarily complicated and can be simplified by making use of the normal kernel and the estimators of Method 2.

Multiplying $\hat{\sigma}_2^2(\cdot)$ by our marginal density estimate $\hat{\pi}(\cdot)$, and differentiating we obtain

$$\frac{d}{dr}\left[\hat{\sigma}_2^2(r)\hat{\pi}(r)\right] = \frac{1}{Th^2\Delta}\sum_{t=1}^{T-1}K'\left(\frac{r-r_{t\Delta}}{h}\right)(r_{(t+1)\Delta} - r_{t\Delta})^2 \qquad (5.24)$$

because $\frac{dK(x/h)}{dx} = h^{-1}K'(x/h)$. Now, using (5.8) and $K'(x) = -xK(x)$, we have

$$\hat{\mu}_3(r) = \frac{1}{2\hat{\pi}(r)}\frac{d}{dr}\left[\hat{\sigma}_2^2(r)\hat{\pi}(r)\right] \qquad (5.25)$$

$$= \frac{1}{2\hat{\pi}(r)\Delta Th^2\sqrt{2\pi}h}\sum_{t=1}^{T-1}e^{-\frac{(r-r_{t\Delta})^2}{2h^2}}(r_{t\Delta}-r)\left(r_{(t+1)\Delta}-r_{t\Delta}\right)^2.$$

As can be seen from (5.25) with (5.23), the form of $\hat{\mu}_3(r)$ is simpler than that of $\hat{\mu}_{JK}(r)$. Now to estimate the diffusion function, we utilize (5.9) and $\hat{\mu}_2(\cdot)$, as well as the information contained in the marginal density, $\hat{\pi}(\cdot)$. So

$$\hat{\sigma}_3^2(r) = \frac{2}{\hat{\pi}(r)}\int_0^r \hat{\mu}_2(u)\hat{\pi}(u)du \qquad (5.26)$$

$$= \frac{2}{\hat{\pi}(r)\Delta Th}\sum_{t=1}^{T-1}[r_{(t+1)\Delta}-r_{t\Delta}]\int_0^r K\left(\frac{u-r_{t\Delta}}{h}\right)du$$

$$= \frac{2}{\hat{\pi}(r)\Delta T}\sum_{t=1}^{T-1}\left[\Phi\left(\frac{r-r_{t\Delta}}{h}\right)-\Phi\left(-\frac{r_{t\Delta}}{h}\right)\right]\cdot[r_{(t+1)\Delta}-r_{t\Delta}].$$

Equations (5.25) and (5.26) provide some explicit and computationally straightforward estimators for $\mu(r)$ and $\sigma^2(r)$. Let $\hat{m}_3(r) = \hat{\mu}_3(r)\hat{\pi}(r)$ and $\hat{V}_3(r) = \hat{\sigma}_3^2(r)\hat{\pi}(r)$. We now have the following propositions, and their proofs are relegated to Section 5.5.

Theorem 5.4. *Assume that Assumptions 5.1–5.3 listed in Section 5.5* *below hold. Then*

$$E\left\{\int [\widehat{m}_3(r) - m(r)]^2\, dr\right\} = O\left(\frac{1}{Th}\right) + O(h^4)$$

$$+\frac{\Delta^2}{4Th^3} \cdot \frac{1}{4\sqrt{\pi}} \int \left(\mu^4(r) + \frac{3\sigma^4(r)}{\Delta^2} + \frac{6\mu^2(r)\sigma^2(r)}{\Delta}\right) \pi(r)dr$$

$$+\int \left(\frac{\Delta}{2}\frac{d}{dr}[\mu^2(r)\pi(r)] + \frac{h^2}{4}\frac{d^3}{dr^3}[\Delta\mu^2(r)\pi(r) + \sigma^2(r)\pi(r)]\right)^2 dr.$$

(i) *When Δ is fixed but $h = d\,T^{-\frac{1}{5}}$ for some $d > 0$, we have*

$$E\left\{\int [\widehat{m}_3(r) - m(r)]^2\, dr\right\} = C_{10}T^{-\frac{2}{5}} + C_{11}\Delta^2 + o\left(T^{-\frac{2}{5}}\right),$$

where both $C_{10} > 0$ and $C_{11} > 0$ are constants.

(ii) *When $\Delta = c\,h^2$ and $h = d\,T^{-\frac{1}{7}}$ for some $c > 0$ and $d > 0$, we have*

$$E\left\{\int [\widehat{m}_3(r) - m(r)]^2\, dr\right\} = C_{12}T^{-\frac{4}{7}} + o\left(T^{-\frac{2}{5}}\right), \qquad (5.27)$$

where $C_{12} > 0$ is a constant.

Theorem 5.5. *Assume that Assumptions 5.1–5.3 listed in Section 5.5 below hold. Then*

$$E\left\{\int \left[\widehat{V}_3(r) - V(r)\right]^2 dr\right\} = h^4 \int \left(\int_0^r \frac{d^2}{dx^2}[\mu(x)\pi(x)]dx\right)^2 dr$$

$$+\frac{4}{T} \cdot \int \left(\int \left[\mu^2(s) + \frac{\sigma^2(s)}{\Delta}\right]^2 \Pi^2(r, s)\pi(s)ds\right) dr + o(h^4),$$

where $\Pi(r, s) = \Phi\left(\frac{r-s}{h}\right) - \Phi\left(-\frac{s}{h}\right).$

(i) *When Δ is fixed but $h = d\,T^{-\frac{1}{5}}$ for some $d > 0$, we have*

$$E\left\{\int \left[\widehat{V}_3(r) - V(r)\right]^2 dr\right\} = C_{13}T^{-\frac{4}{5}} + o\left(T^{-\frac{4}{5}}\right), \qquad (5.28)$$

where $C_{13} > 0$ is a constant.

(ii) *When $\Delta = c\,h^2$ and $h = d\,T^{-\frac{1}{7}}$ for some $c > 0$ and $d > 0$, we have*

$$E\left\{\int \left[\widehat{V}_3(r) - V(r)\right]^2 dr\right\} = C_{14}T^{-\frac{3}{7}} + o\left(T^{-\frac{3}{7}}\right), \qquad (5.29)$$

where $C_{14} > 0$ is a constant.

Overall, we suggest using

- $h = d\,T^{-\frac{1}{5}}$ when Δ is fixed, and $\Delta = c\,h^2$ and $h = d\,T^{-\frac{1}{7}}$ when Δ is varied according to T.

We may also need to consider the pairs $(\widehat{m}_2, \widehat{V}_3)$ and $(\widehat{m}_3, \widehat{V}_2)$ separately, since $\widehat{V}_3$ is constructed using $\widehat{m}_2$, and $\widehat{m}_3$ is based on $\widehat{V}_2$. In this case, we should suggest using

- $\Delta = c\, h$ and $h = d\, T^{-\frac{1}{6}}$ for Theorems 5.2 and 5.5. In this case, the resulting rates are

$$E\left\{\int [\widehat{m}_2(r) - m(r)]^2\, dr\right\} = C_1(1 + o(1))T^{-\frac{4}{6}},$$

$$E\left\{\int \left[\widehat{V}_3(r) - V(r)\right]^2 dr\right\} = C_2(1 + o(1))T^{-\frac{4}{6}}. \qquad (5.30)$$

- $\Delta = c\, h^2$ and $h = d\, T^{-\frac{1}{9}}$ for Theorem 5.3, and the resulting rate is

$$E\left\{\int \left[\widehat{V}_2(r) - V(r)\right]^2 dr\right\} = C_3(1 + o(1))T^{-\frac{4}{9}}. \qquad (5.31)$$

- $\Delta = c\, h^2$ and $h = d\, T^{-\frac{1}{7}}$ for Theorem 5.4, and the resulting rate is

$$E\left\{\int [\widehat{m}_3(r) - m(r)]^2\, dr\right\} = C_4(1 + o(1))T^{-\frac{4}{7}}. \qquad (5.32)$$

This shows that we may consider linking Δ with h or vice versa for the case where Δ is varied according to T when applying a nonparametric kernel method to estimate a discretized version of a continuous–time model.

We suggest using the pair $(\widehat{\mu}_2(r), \widehat{\sigma}_3^2(r))$ in theory. The empirical comparisons in Section 5.4 show that the pairs $(\widehat{\mu}_2(r), \widehat{\sigma}_2^2(r))$ and $(\widehat{\mu}_3(r), \widehat{\sigma}_3^2(r))$ are both appropriate for the two sets of data. By considering both the theoretical properties and empirical comparisons of the proposed estimators, however, we would suggest using the pair $(\widehat{\mu}_2(r), \widehat{\sigma}_3^2(r))$ for the two sets of data. We also use $h = c \cdot T^{-\frac{1}{5}}$ with c to be specified later in the empirical comparisons below mainly because real data are normally available in a discrete form and thus Δ is fixed in practice.

5.3 Semiparametric specification

In this section, we discuss two test statistics for parametric specification of both the drift and diffusion functions. Both the theory and the implementation have been established in Casas and Gao (2005) and Arapis and Gao (2006). To make a relevant application of the proposed tests, we then examine the large and finite sample performance of a test for linearity in the drift function through using the two financial data sets.

5.3.1 Specification of diffusion function

Throughout this section, we consider a semiparametric diffusion model of the form

$$dr_t = \mu(r_t, \theta)dt + \sigma(r_t)dB_t, \tag{5.33}$$

where $\mu(r, \theta)$ is a known parametric function indexed by a vector of unknown parameters, $\theta \in \Theta$ (a parameter space), and $\sigma(r)$ is an unknown but sufficiently smooth function. As pointed out in Kristensen (2004), there is sufficient evidence that the assumption of a parametric form for the drift function is not unreasonable. In addition, Arapis and Gao (2006) have shown that when the drift function is unknown nonparametrically, the drift function may be specified parametrically without knowing the form of $\sigma(\cdot)$.

Let $Y_t = \frac{r_{(t+1)\Delta} - r_{t\Delta}}{\Delta}$, $X_t = r_{t\Delta}$, $f(x, \theta) = \mu(x, \theta)$ and $g(x) = \Delta^{-1}\sigma^2(x)$. Model (5.2) suggests approximating model (5.33) by a nonparametric autoregressive model of the form

$$Y_t = f(X_t, \theta) + \epsilon_t \quad \text{with} \quad \epsilon_t = \sqrt{g(X_t)}\, e_t, \tag{5.34}$$

where $\{e_t\}$ is a sequence of independent N(0,1) errors and independent of $\{X_s\}$ for all $s \leq t$. So $E[e_t|X_t] = E[e_t] = 0$ and $\mathrm{var}[e_t|X_t] = \mathrm{var}[e_t] = 1$.

The main interest of this section is to test

$$\begin{aligned}\mathcal{H}_{051} : g(x) &= g(x, \vartheta_0) \text{ versus}\\ \mathcal{H}_{151} : g(x) &= g(x, \vartheta_1) + C_{51} \cdot D_{51}(x)\end{aligned} \tag{5.35}$$

for all $x \in \mathbf{R}^1$ and some $\vartheta_0, \vartheta_1 \in \Theta$, where both ϑ_0 and ϑ_1 are to be chosen, Θ is a parameter space, C_{51} is a sequence of real numbers, and $D_{51}(x)$ is a specified and smooth function. Note that ϑ_0 may be different from the true value, θ_0, of θ involved in the drift function.

In order to construct our test for $\mathcal{H}_{051}$, we use (5.34) to formulate a regression model of the form

$$\epsilon_t^2 = g(X_t) + \eta_t, \tag{5.36}$$

where the error process $\eta_t = g(X_t)(e_t^2 - 1)$ is of the following properties: under $\mathcal{H}_{051}$

$$E[\eta_t|X_t] = 0 \quad \text{and} \quad E[\eta_t^2|X_t] = 2g^2(X_t, \vartheta_0). \tag{5.37}$$

In general, for any $k \geq 1$ we have under $\mathcal{H}_{051}$

$$E\left[\eta_t^k|X_t\right] = E\left[(e_t^2 - 1)^k\right]\, g^k(X_t, \vartheta_0) \equiv c_k g^k(X_t, \vartheta_0), \tag{5.38}$$

where $c_k = E\left[(e_t^2 - 1)^k\right]$ is a known value for each k using the fact that $e_t \sim N(0, 1)$ has all known moments. This implies that all higher–order conditional moments of $\{\eta_t\}$ will be specified if the second conditional moment of $\{\eta_t\}$ is specified.

Since model (5.34) is a special case of model (3.32), we suggest using a test statistic of the form

$$L_{51}(h) = \widehat{L}_{0T}(h) \qquad (5.39)$$

as defined in (3.36) in Chapter 3. As a direct application, Theorem 3.6 implies an asymptotically normal test for $\mathcal{H}_{051}$.

5.3.2 Specification of drift function

Throughout this section, we consider a semiparametric diffusion model of the form

$$dr_t = \mu(r_t)dt + \sigma(r_t, \vartheta)dB_t, \qquad (5.40)$$

where $\sigma(r, \vartheta)$ is a positive parametric function indexed by a vector of unknown parameters, $\vartheta \in \Theta$ (a parameter space), and $\mu(r)$ is an unknown but sufficiently smooth function. As pointed out in existing studies, such as Kristensen (2004), there is some evidence that the assumption of a parametric form for the diffusion function is also reasonable in such cases where the diffusion function is already pre–specified, the main interest is, for example, to specify whether the drift function should be linear or quadratic. In Arapis and Gao (2006), the authors have discussed how to specify the drift function parametrically while the diffusion function is allowed to be unknown nonparametrically.

Similarly to model (5.34), we suggest approximating model (5.40) by a semparametric autoregressive model of the form

$$Y_t = f(X_t) + \sqrt{g(X_t, \vartheta)}\, e_t, \qquad (5.41)$$

where $f(X_t) = \mu(X_t)$, $g(X_t, \vartheta) = \Delta^{-1}\sigma^2(X_t, \vartheta)$, and $\{e_t\}$ is a sequence of independent Normal errors with $E[e_t|X_t] = E[e_t] = 0$ and $\mathrm{var}[e_t|X_t] = \mathrm{var}[e_t] = 1$.

Our interest is then to test

$$
\begin{aligned}
\mathcal{H}_{052} :\ f(x) &= f(x, \theta_0) \ \text{ versus} \\
\mathcal{H}_{152} :\ f(x) &= f(x, \theta_1) + C_{52} \cdot D_{52}(x)
\end{aligned}
\qquad (5.42)
$$

for all $x \in \mathbf{R}^1$ and some θ_0, $\theta_1 \in \Theta$, where Θ is a parameter space, C_{52} is a sequence of real numbers, and $D_{52}(x)$ is a specified and smooth function. Note that θ_0 may be different from the true value, ϑ_0, of ϑ involved in the diffusion function.

Similarly to the construction of $L_{51}(h)$, we propose using a normalized version of the form

$$L_{52}(h) = \frac{\sum_{s=1}^{T} \sum_{t=1, \neq t}^{T} \widehat{\epsilon}_s \, K\left(\frac{X_s - X_t}{h}\right) \widehat{\epsilon}_t}{\widehat{\sigma}_{52}}, \qquad (5.43)$$

where $\widehat{\epsilon}_t = Y_t - f(X_t, \widehat{\theta}_0)$ with $\widehat{\theta}_0$ being a $\sqrt{T}$–consistent estimator of θ_0, and $\widehat{\sigma}_{52}^2 = 2\widehat{\nu}_2^2 \int K^2(u) du$ with $\widehat{\nu}_2 = \frac{1}{T} \sum_{t=1}^{T} g(X_t, \widehat{\vartheta})$, in which $\widehat{\vartheta}$ is a $\sqrt{T}$-consistent estimator of ϑ. Since the diffusion function is prespecified parametrically, we need not involve any nonparametric estimator in $\widehat{\sigma}_{52}^2$.

Similarly to Theorem 3.6, we may show that $L_{52}(h)$ is an asymptotically normal test. Such details have been given in Casas and Gao (2005). Since the details are very analogous, we do not wish to repeat them. Instead, we focus on testing for linearity in the drift function in the following section.

5.3.3 Testing for linearity in the drift

To formally determine whether the assumption on linearity in the drift in Method 1 is appropriate for a given set of data, we consider testing

$$\begin{aligned} \mathcal{H}_{053} : \mu(r) &= \mu(r; \theta_0) = \beta_0(\alpha_0 - r) \quad \text{versus} \\ \mathcal{H}_{153} : \mu(r) &= \mu(r; \theta_1) = \gamma_1 + \beta_1(\alpha_1 - r)r \end{aligned} \qquad (5.44)$$

for all $r \in \mathbf{R}^+ = (0, \infty)$ and some $\theta_0 = (\alpha_0, \beta_0) \in \Theta$, where $\theta_1 = (\alpha_1, \beta_1, \gamma_1) \in \Theta$ (a parameter space in $\mathbf{R}^3$) is chosen such that the alternative is different from the null. Equation (5.44) shows that we are interested in testing for a mean–reverting drift versus a quadratic drift.

We approximate the semiparametric continuous–time diffusion model $dr_t = \beta(\alpha - r_t) dt + \sigma(r_t) dB_t$ by a semiparametric time series model of the form

$$Y_t = \beta(\alpha - X_t) + \sigma(X_t) e_t, \qquad (5.45)$$

where $X_t = r_{t\Delta}$, $Y_t = \frac{X_{t+1} - X_t}{\Delta}$, $\sigma(\cdot) > 0$ is unknown nonparametrically, and $e_t = \frac{B_{(t+1)\Delta} - B_{t\Delta}}{\Delta} \sim N\left(0, \Delta^{-1}\right)$.

We initially suggest a specification test of the form

$$L_{53}(h) = \frac{\sum_{s=1}^{T} \sum_{t=1, t\neq s}^{T} \widehat{\epsilon}_s \, K\left(\frac{X_s - X_t}{h}\right) \widehat{\epsilon}_t}{\sqrt{2 \sum_{s=1}^{T} \sum_{t=1}^{T} \widehat{\epsilon}_s^2 \, K^2\left(\frac{X_s - X_t}{h}\right) \widehat{\epsilon}_t^2}}, \qquad (5.46)$$

where $K(\cdot)$ is the standard normal density function, h is the bandwidth parameter, and $\widehat{\epsilon}_t = Y_t - \widehat{\mu}_1(X_t)$.

As can be seen from the proof of Lemma 5.1 below, it may be shown that under $\mathcal{H}_{053}$

$$L_{53}(h) = \frac{\sum_{s=1}^{T}\sum_{t=1,t\neq s}^{T} \epsilon_s \, K\left(\frac{X_s-X_t}{h}\right) \epsilon_t}{\sqrt{2\sum_{s=1}^{T}\sum_{t=1}^{T} \epsilon_s^2 \, K^2\left(\frac{X_s-X_t}{h}\right) \epsilon_t^2}} + o_P(1) \qquad (5.47)$$

$$= \frac{\sum_{s=1}^{T}\sum_{t=1,t\neq s}^{T} u_s \, \sigma(X_s)K\left(\frac{X_s-X_t}{h}\right) \sigma(X_t) \, u_t}{\sqrt{2\sum_{s=1}^{T}\sum_{t=1}^{T} u_s^2 \, \sigma^2(X_s)K^2\left(\frac{X_s-X_t}{h}\right) \sigma^2(X_t) \, u_t^2}} + o_P(1)$$

for sufficiently large T, where $u_t = \sqrt{\Delta}e_t \sim N(0,1)$ is independent of X_s for all $s \leq t$.

As argued in Li and Wang (1998) and Li (1999), the first part of Equation (5.47) shows that the test statistic $L_{53}(h)$ has the main feature that it appears to be more straightforward computationally than other kernel–based tests (see Härdle and Mammen 1993; Hjellvik and Tjøstheim 1995; Hjellvik, Yao and Tjøstheim 1998), since it is not required to get a consistent estimator of the conditional variance involved. Furthermore, the second part of Equation (5.47) shows that the leading term of $L_{53}(h)$ involves an error process $\{u_t\}$ independent of Δ. In addition, it has been shown in Lemma 5.1 below that as $T \rightarrow \infty$, $L_{53}(h)$ converges in distribution to $N(0,1)$ regardless of whether Δ is fixed or varied according to T. This implies that in theory the applicability of the test for testing the drift depends on neither the structure of the conditional variance nor the choice of Δ. In practice, $L_{53}(h)$ is computed using monthly, weekly, daily, or higher frequency observations. Therefore, it is appropriate to apply the test to our case study once the bandwidth is appropriately chosen.

It follows from Theorem A.2 in the appendix that $L_{53}(h)$ converges in distribution to the standard normality when $T \rightarrow \infty$. Our experience and others show that the finite sample performance of $L_{53}(h)$ is not good in particular when h is chosen based on an optimal estimation procedure, such as the cross–validation criterion. The main reasons are as follows: (a) the use of an estimation-based optimal value may not be optimal for testing purposes; and (b) the rate of convergence of $L_{53}(h)$ to the asymptotic normality is quite slow even when $\{e_t\}$ is now a sequence of independent and normally distributed errors. With respect to the choice of a suitable bandwidth for testing purposes, we should propose choosing a suitable bandwidth based on the assessment of both the size and power functions of $L_{53}(h)$. Consequently, the issue of choosing Δ may also be addressed when using the discretized version (5.2) for continuous–time model specification.

Since we have not been able to solve such a choice problem for the continuous–time diffusion case, we instead propose using the following two schemes to improve the finite sample performance of $L_{53}(h)$. We first establish an adaptive version of $L_{53}(h)$ over a set of all possible bandwidth values. Second, we use a simulated critical value for computing the size and power values of the adaptive version of $L_{53}(h)$ instead of using an asymptotic value of $l_{0.05} = 1.645$ at the 5% level. To the best of our knowledge, both the schemes are novel in this kind of testing for linearity in the drift under such a semiparametric setting.

We then propose using an adaptive test of the form

$$L^* = \max_{h \in H_T} L_{53}(h), \tag{5.48}$$

where $H_T = \{h = h_{\max} a^k : h \geq h_{\min}, \ k = 0, 1, 2, \ldots\}$, in which $0 < h_{\min} < h_{\max}$, and $0 < a < 1$. Let J_T denote the number of elements of H_T. In this case, $J_T \leq \log_{1/a}(h_{\max}/h_{\min})$.

Simulation Scheme: We now discuss how to obtain a simulated critical value for L^*. The exact α–level critical value, l_{α}^e $(0 < \alpha < 1)$ is the $1 - \alpha$ quantile of the exact finite–sample distribution of L^*. Because l_{α}^e may not be evaluated in practice, we therefore suggest choosing a simulated α–level critical value, l_{α}^*, by using the following simulation procedure:

1. For each $t = 1, 2, \ldots, T$, generate $Y_t^* = \widehat{\mu}_1(X_t) + \widehat{\sigma}_1(X_t)e_t^*$, where $\{e_t^*\}$ is sampled randomly from $N(0, \Delta^{-1})$ for Δ to be specified as either $\Delta = \frac{20}{250}$ for the monthly data or $\Delta = \frac{1}{250}$ for the daily data, which $\widehat{\mu}_1(\cdot)$ and $\widehat{\sigma}_1(\cdot)$ are as defined in (5.10) and (5.11), respectively. In practice, a kind of truncation procedure may be needed to ensure the positivity of $\widehat{\sigma}_1(\cdot)$.

2. Use the data set $\{Y_t^* : t = 1, 2, \ldots, T\}$ to re-estimate θ_0. Denote the resulting estimate by $\widehat{\theta}^*$. Compute the statistic $\widehat{L}^*$ that is obtained by replacing Y_t and $\widehat{\theta}$ with Y_t^* and $\widehat{\theta}^*$ on the right–hand side of (5.48).

3. Repeat the above steps M times and produce M versions of $\widehat{L}^*$ denoted by $\widehat{L}_m^*$ for $m = 1, 2, \ldots, M$. Use the M values of $\widehat{L}_m^*$ to construct their empirical bootstrap distribution function, that is, $F^*(u) = \frac{1}{M} \sum_{m=1}^{M} I(\widehat{L}_m^* \leq u)$. Let l_{α}^* be the $1 - \alpha$ quantile of the empirical bootstrap distribution and then estimate l_{α}^e by l_{α}^*.

We now state the following results, and their proofs are relegated to Section 5.5.

Theorem 5.6. *Assume that Assumptions 5.1(i), 5.2 and 5.4 listed in Section 5.5 below hold. Then under $\mathcal{H}_{053}$*

$$\lim_{T \to \infty} P(L^* > l_{\alpha}^*) = \alpha.$$

The main result on the behavior of the test statistic L^* under $\mathcal{H}_{053}$ is that l_α is an asymptotically correct α–level critical value under any model in $\mathcal{H}_{053}$.

Theorem 5.7. *Assume that the conditions of Theorem 5.6 listed in Section 5.5 below hold. Then under* $\mathcal{H}_{153}$

$$\lim_{T \to \infty} P(L^* > l_\alpha^*) = 1.$$

Theorem 5.7 shows that a consistent test will reject a false $\mathcal{H}_{053}$ with probability approaching one as $T \to \infty$. It is pointed out that Theorems 5.6 and 5.7 are new in this kind of continuous–time diffusion model specification.

To implement Theorems 5.6 and 5.7 to real data analysis, we need to compute the p–value of the test for each given set of data as follows:

1. For either the Fed rate or the Eurodollar rate data, compute

$$L^* = \max_{h \in H_T} L_{53}(h) \tag{5.49}$$

with $H_T = \{h = h_{\max} a^k : h \geq h_{\min}, \ k = 0, 1, 2, \dots\}$, where $T^{-\frac{1}{5}} = h_{\min} < h_{\max} = 1.1 \, (\log\log T)^{-1}$, and $a = 0.8$ based on preliminary calculations of the size and power values of $L_{53}(h)$ for a range of bandwidth values.

2. Compute $\widehat{\epsilon}_t = Y_t - \widehat{\mu}_1(X_t)$ and then generate a sequence of bootstrap resamples $\{\widehat{\epsilon}_t^*\}$ using the wild bootstrap method (see Härdle and Mammen 1993; Li and Wang 1998) from $\{\widehat{\epsilon}_t\}$.

3. Generate $\widehat{Y}_t^* = \widehat{\mu}_1(X_t) + \widehat{\epsilon}_t^*$. Compute the corresponding version $\widehat{L}^*$ of L^* based on $\{\widehat{Y}_t^*\}$.

4. Repeat the above steps N times to find the bootstrap distribution of $\widehat{L}^*$ and then compute the proportion that $L^* < \widehat{L}^*$. This proportion is a simulated p–value of L^*.

With the three methods now constructed, it is of interest to see how they compare, and more precisely, how the restriction placed on the drift function affects the estimation of the diffusion. A number of studies note that the prices of derivatives are crucially dependent on the specification of the diffusion function (see Aït-Sahalia 1996a), therefore qualifying parametric restrictions on the drift function. The test statistic L^* is also applied to formally test linearity in the drift using the simulated p–value.

5.4 Empirical comparisons

5.4.1 The data

We now apply the three pairs of estimators constructed previously to two different financial data. A conclusion regarding which method best fits each data set will be offered. Also suggested here is an optimal bandwidth, based on both the theoretical properties of the MISEs in Theorems 5.1–5.5 and a comparison of a number of common forms used in the literature.

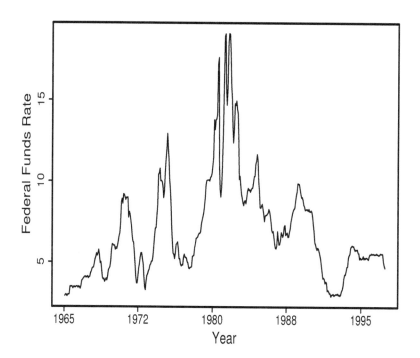

Figure 5.1 *Three-month T-Bill rate, January 1963 to December 1998.*

To analyze the effect the sampling frequency (interval) has on the results, we use both monthly (low frequency) and daily (high frequency) sampled data. The three-month Treasury Bill rate data set given in Figure 5.1 is sampled monthly over the period from January 1963 to December 1998, providing 432 observations (i.e. $T = 432$; source: H–15 Federal Reserve Statistical Release). The number of working days in a year (excluding weekends and public holidays) is assumed to be 250 (and 20 working days per month). This gives $\Delta = \frac{20}{250}$. Chan *et al.* (1992) offer evidence that the Fed rates are stationary by showing that the autocorrelations of month–to–month changes are neither large nor consistently positive or negative.

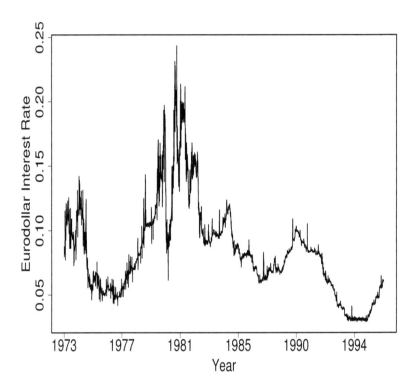

Figure 5.2 *Seven-Day Eurodollar Deposit rate, June 1, 1973 to February 25, 1995.*

The second data set used in this analysis to compare and contrast the primary results is the high frequency seven–day Eurodollar deposit rate. The data are sampled daily from June 1, 1973 to February 25, 1995. This provides us with $T = 5505$ observations. Just as for the Fed data, holidays have not been treated and Monday is taken as the first day after

Friday as there are no obvious weekend effects (Aït-Sahalia 1996b). Thus, our sampling interval $\Delta = \frac{1}{250}$. The data are plotted in Figure 5.2.

For the stationarity test of the data sets, Aït-Sahalia (1996a) and Jiang (1998) rejected the null hypothesis of nonstationarity on the respective Eurodollarand Fed data based on results of an augmented Dickey–Fuller nonstationarity test.

5.4.2 Bandwidth choice

The choice of bandwidth is critical in any application of nonparametric kernel density and regression estimation. Theorems 5.1–5.5 provide some kind of guidance on how to choose the bandwidth in practice. Overall, we suggest using $h = d\,T^{-\frac{1}{5}}$ when Δ is fixed and $\Delta = c\,h^2$ and $h = d\,T^{-\frac{1}{7}}$ when $\Delta \to 0$. As we deal with the fixed Δ in our empirical study, the forms of the bandwidth selectors used are listed below:

$$h_1 = s_d \times T^{-\frac{1}{5}}, \ h_2 = \frac{1}{10} \times T^{-\frac{1}{5}}, \ h_3 = \frac{1}{4} \times T^{-\frac{1}{5}}, \ h_4 = 1.06 \times s_d \times T^{-\frac{1}{5}},$$

where T is the number of observations and s_d is the standard deviation of the data. Thus h_1 and h_4 can, in this sense, be regarded as "data-driven" bandwidth choices. Pritsker (1998) stated that h_4 is the MISE–minimizing bandwidth assuming the data came from a normal distribution with variance s_d^2. As can be seen from Theorems 5.1–5.5, the second and third bandwidths can be written as $h = c \cdot T^{-\frac{1}{5}}$, where c is a constant chosen to minimize the asymptotic MISE of the estimator involved. Our detailed graphical comparison shows that h_2 is the optimal one in terms of not only providing a smooth and informative marginal density estimate (see Figure 5.3), but also possessing the greatest consistency between the three drift and diffusion estimators (see Figure 5.4 for the Fed rate and Figure 5.5 for the Euro data). In addition to such key figures, we have also produced some other figures for both the drift and diffusion estimators based on Methods 1–3. For the Euro data, we also borrowed bandwidth choices used by Aït-Sahalia (1996a, 1996b), as he has also used this data set.

5.4.3 Results and comparisons for the Fed data

We "plugged-in" the bandwidths h_1, h_2, h_3, h_4 of $0.00949, 0.0297, 0.0743$ and 0.01, respectively, and estimated the marginal density, drift and diffusion functions for the Fed data. It was found that the optimal bandwidth refers to h_2 (0.0297). The density estimate produced for h_2 shown

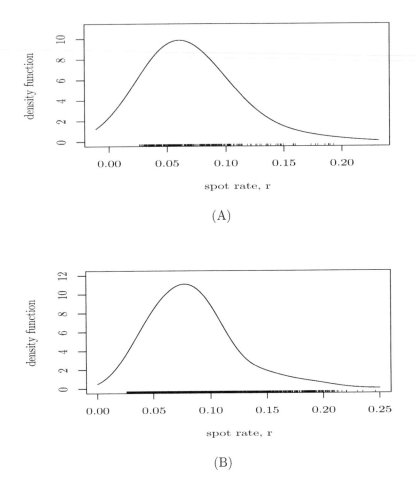

Figure 5.3 *A: Nonparametric kernel density estimator for the Fed data with*
$h_2 = 0.0297$. *B: Nonparametric kernel density estimator for the Euro data*
with $h_2 = 0.01786$.

in Figure 5.3(A) appears to contain sufficient information. It is apparent
with this choice of bandwidth, even though the high rate period of 1980–
82 is included in the sample, the amount of information retained has
produced a less accentuated right tail. Its shape and symmetry about
0.055 closely resembles that of a Gaussian density. The densities produced with the smaller bandwidths were overly informative while larger
bandwidths resulted in smooth quadratic like curves.

In Figures 5.4–5.13, the pairs of the estimators from the top to the bottom correspond to $(\widehat{\mu}_2, \widehat{\sigma}_2^2)$, $(\widehat{\mu}_3, \widehat{\sigma}_3^2)$ and $(\widehat{\mu}_1, \widehat{\sigma}_1^2)$.

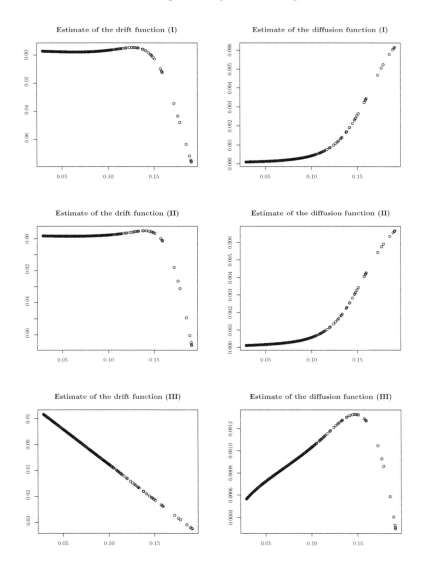

Figure 5.4 *Nonparametric drift and diffusion estimators for the Fed data with* $h_2 = 0.01786$.

Comparisons of the drift and diffusion estimators give similar results. The three drift and diffusion estimators constructed using our optimal

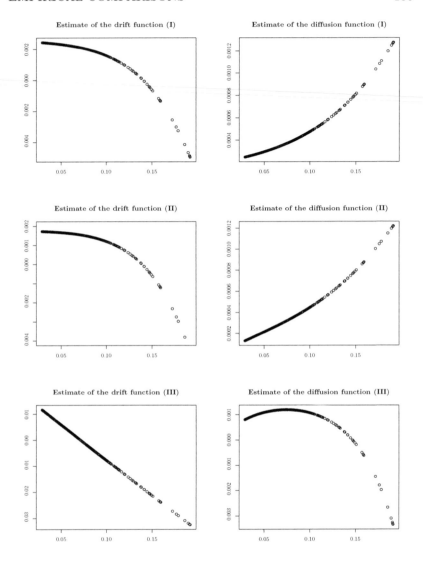

Figure 5.5 *Nonparametric drift and diffusion estimators for the Fed data with* $h_3 = 0.0743$.

bandwidth choice are superimposed for comparative purposes in Figure 5.7. The best estimators for the Fed data are given in Figure 5.7. The drift functions $\widehat{\mu}_2(\cdot)$ and $\widehat{\mu}_3(\cdot)$ inherit similar nonlinearity for interest rates over the entire range of r. The best linear mean-reverting drift

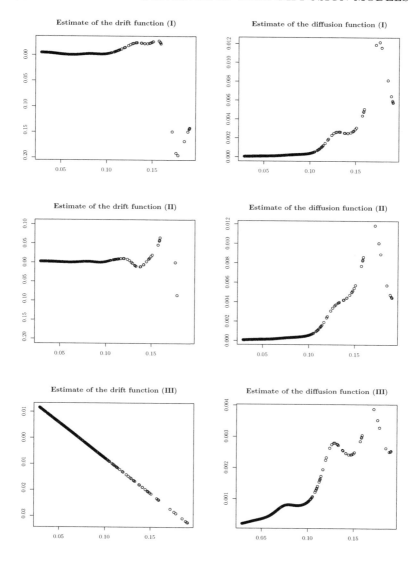

Figure 5.6 *Nonparametric drift and diffusion estimators for the Fed data with* $h_1 = 0.00949$.

estimate is plotted in the bottom of Figure 5.4 and then Figure 5.7. The ordinary least squares method gave estimates for the parameters α and β of $\widehat{\mu}_1(\cdot)$ of 0.07170 and 0.2721, respectively.

Looking now at the diffusion estimators we see $\widehat{\sigma}_2^2(\cdot)$ and $\widehat{\sigma}_3^2(\cdot)$ especially

Drift functions (I),(II) & (III)

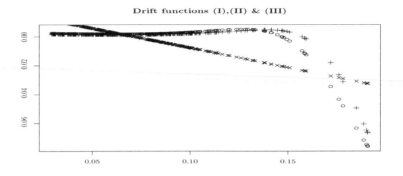

Diffusion functions (I),(II) & (III)

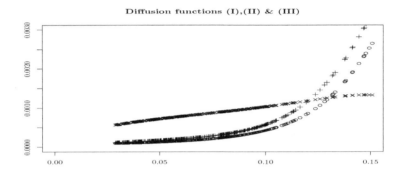

Figure 5.7 *The best estimators for the Fed data with $h_2 = 0.0297$. The $(\times)$ refers to $\widehat{\mu}_1$ and $\widehat{\sigma}_1^2$, $(\circ)$ to $\widehat{\mu}_2$ and $\widehat{\sigma}_2^2$ and $(+)$ to $\widehat{\mu}_3$ and $\widehat{\sigma}_3^2$.*

are very similar. They closely resemble one another over the entire range of r in both shape and magnitude (see Figures 5.4–5.7). The curvature of $\widehat{\sigma}_2^2(\cdot)$ and $\widehat{\sigma}_3^2(\cdot)$ is close to that of a quadratic. This gives some support for the process of Brennan and Schwartz (1980), whose instantaneous variance increased at a rate proportional to r^2, and to Chan *et al.* (1992), who found $\sigma \propto r^{1.49}$.

The best estimator $\widehat{\sigma}_1^2(\cdot)$ (see the bottom of Figure 5.4 and then Figure 5.7) is comparable with $\widehat{\sigma}_2^2(\cdot)$ and $\widehat{\sigma}_3^2(\cdot)$ for low to moderate rates (i.e., rates below 12%). It lies above $\widehat{\sigma}_2^2(\cdot)$ and $\widehat{\sigma}_3^2(\cdot)$ for a greater (negative) mean–reverting force ($\widehat{\mu}_1(\cdot) < \widehat{\mu}_2(\cdot)$, $\widehat{\mu}_3(\cdot)$). It appears to be a linearly increasing function of the level of r (as in Cox, Ingersoll and Ross (CIR) 1985) for rates below 14%, and this is apparent in the bottom of Figure 5.4 and then Figure 5.7.

Given that the two nonparametric drift estimators are unlike the linear

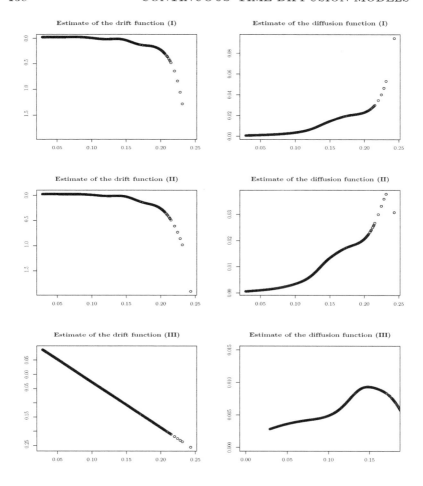

Figure 5.8 *Nonparametric drift and diffusion estimators for the Euro data with* $h_2 = 0.01786$.

mean-reverting specification (with their respective diffusion estimates $\widehat{\sigma}_2^2(\cdot)$ and $\widehat{\sigma}_3^2(\cdot)$ differing from $\widehat{\sigma}_1^2(\cdot)$), we suggest here that the mean–reverting function is not appropriate for these data. Thus, $\widehat{\mu}_1(\cdot)$ does indeed affect the estimation of the diffusion function and hence the pricing of derivative securities. Based on the above, for the monthly sampled Federal funds rate data, we believe that $\widehat{\mu}_1(\cdot)$ imposes an unnecessary restriction that results in the misspecification of the diffusion function. Either of the pair $(\widehat{\mu}_2(\cdot), \widehat{\sigma}_2^2(\cdot))$ or $(\widehat{\mu}_3(\cdot), \widehat{\sigma}_3^2(\cdot))$ is recommended for this set of data.

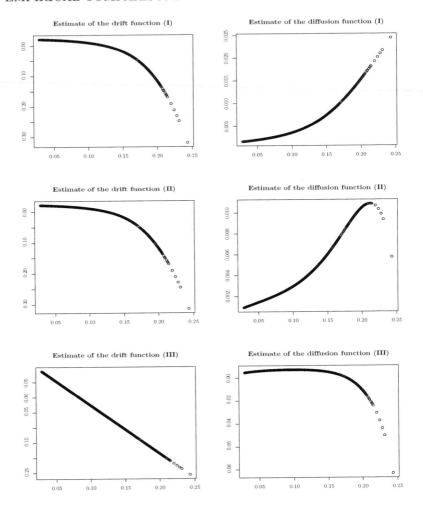

Figure 5.9 *Nonparametric drift and diffusion estimators for the Euro data with* $h_3 = 0.044$.

The specification test L^* proposed in Section 5.3 was then applied in order to formally reject linearity in the drift. The null hypothesis H_0 : $\mu(r) = \beta(\alpha - r)$ of linearity is rejected at the 5% significance level. We obtain a simulated p–value of $p \leq 0.001$, which is much smaller than the 5% significance level.

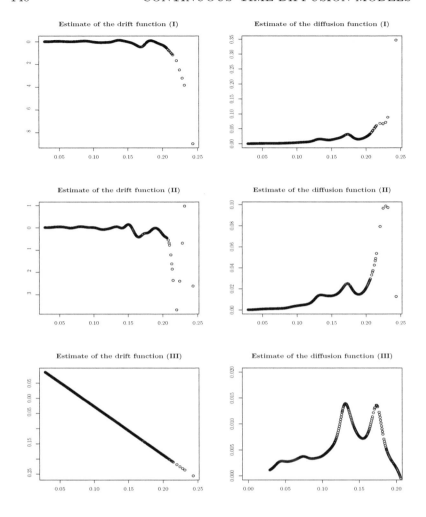

Figure 5.10 *Nonparametric drift and diffusion estimators for the Euro data with $h_1 = 0.006413$.*

5.4.4 Results and comparisons for the Euro data

Now to the Euro data. The forms of $h_1, \ldots, h_4$ were applied to these data. In this case, because we have 5505 observations, the bandwidths were, respectively, $0.006413, 0.01786, 0.044$, and 0.0068. We also consider the bandwidth $h_a = 0.01347$ used by Aït-Sahalia (1996a) for the same data. The best estimators for the Euro data are given in Figures 5.8 and 5.11. Surprisingly, our optimal bandwidth for the Euro data also

Drift functions (I),(II) & (III)

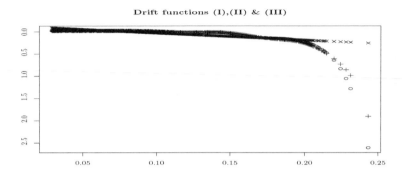

Diffusion functions (I),(II) & (III)

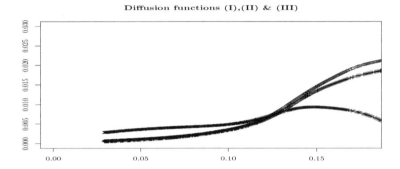

Figure 5.11 *The best estimators for the Euro data with* $h_2 = 0.01786$. *The* $(\times)$ *refers to* $\widehat{\mu}_1$ *and* $\widehat{\sigma}_1^2$, $(\circ)$ *to* $\widehat{\mu}_2$ *and* $\widehat{\sigma}_2^2$ *and* $(+)$ *to* $\widehat{\mu}_3$ *and* $\widehat{\sigma}_3^2$.

corresponds to $h_2 = \frac{1}{10} \times T^{-\frac{1}{5}}$. Similar to the Fed data analysis where we concluded h_1 and h_4 severely undersmooth the density estimate, we infer similar results for the Euro data. With the Euro data consisting of 5505 observations, it is clear we would obtain a much smaller sample variance than the Fed data (consisting of 432 observations). Our best marginal density estimate, drift and diffusion estimators for the Euro data are reported in Figures 5.3(B) and 5.8. The drift and diffusion estimators are superimposed for comparative purposes in Figure 5.11. It is apparent that the two unrestricted drift estimators, $\widehat{\mu}_2(\cdot)$ and $\widehat{\mu}_3(\cdot)$, inherit very similar nonlinearity over the entire range of r (see Figure 5.11). Both estimators seem to exhibit mean reversion for $r > 15\%$, while our linear mean–reverting drift estimator $\widehat{\mu}_1(\cdot)$ (see also Figure 5.11) is unexpectedly comparable with $\widehat{\mu}_2(\cdot)$ and $\widehat{\mu}_3(\cdot)$ for $r < 20\%$. The diffusion functions constructed using the unrestricted drift closely resemble one another and are practically indistinguishable for the entire range of r.

They both increase somewhat linearly for $r < 11\%$, both increase at a greater rate than r for $r > 11\%$ and possess a "hump" at $r = 15\%$ where the instantaneous variance jumps (see Figure 5.11).

For the Euro data, the best OLS estimates of α and β are 0.08308 and 1.596, respectively, which are analogous to the first step OLS estimates computed by Aït-Sahalia (1996a). We see the Euro data have stronger mean-reversion than the Fed data ($\beta = 0.2721$), which is most likely the result of more frequent sampling. Aït-Sahalia (1996a) also found β to be larger for shorter-maturity proxies (seven-day Eurodollar versus three-month T-bill). We see from Figure 5.11 that the similarity of the three drift estimators may suggest the mean-reverting specification for drift is applicable (at least for $r < 20\%$). The similarity of the two diffusion functions $\hat{\sigma}_2^2(\cdot)$ and $\hat{\sigma}_3^2(\cdot)$ and the deviation of $\hat{\sigma}_1^2(\cdot)$ from these two estimators may, however, suggest otherwise.

The proposed test L^* was run on the data. Our detailed simulation returns a simulated p–value of $p \leq 0.001$, which directs us to strongly reject the null hypothesis of linearity at the 5% significance level. A likely explanation for this result is that as we have a long and frequently sampled data set, the use of even the slightest deviant from the actual drift will result in a compounded error effect or deviation of the specified model from the actual process. Thus, we suggest the mean-reverting drift function specification is not appropriate for high frequency data (more strongly than for the monthly sampled Fed data). To determine whether the high rate period of 1980–82 was responsible for the strong rejection of linearity, we also ran the test on the sub–sample and calculated p–value. The result suggests that linearity in the drift is also rejected for the subsample. Not only did we run the linearity test on the subsample of the Euro data, but we also estimated the drift and diffusion estimators for this set. Here, we "plugged-in" the bandwidth value of $h_s = 0.01653$, which Aït-Sahalia (1996a) reported was optimal for this subsample. The resulting drift and diffusion estimators are given in Figure 5.12. The two unrestricted drift estimators exhibit similar nonlinearities for $r < 10\%$ with mean-reversion for $r > 10\%$ while their corresponding diffusion estimators resemble the quadratic diffusion specification of Brennan and Schwartz (1980). The diffusion estimator $\hat{\sigma}_1^2(\cdot)$ appears to be comparable with the constant volatility specification of Vasicek (1977) for $r < 12\%$. Such a difference in form is evidence against the linear mean–reverting drift function.

Function estimates computed using the borrowed bandwidth of $h_a = 0.01347$ give results that are slightly suboptimal in our opinion (see Figure 5.13). A comparison of our diffusion estimator $\hat{\sigma}_1^2(\cdot)$ with the diffusion estimator of Aït-Sahalia (1996a) (see Figure 4 of his paper) for the

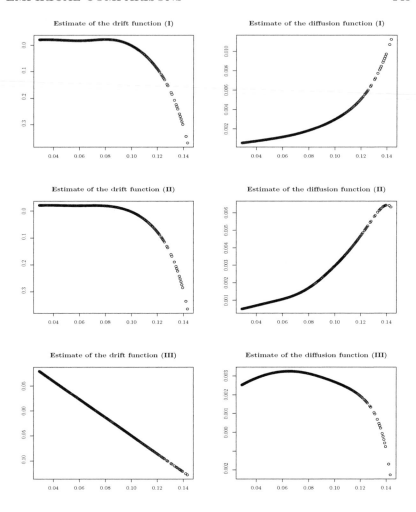

Figure 5.12 *Nonparametric drift and diffusion estimators for the Euro sub-sample with $h_s = 0.01653$.*

same data set suggests that the estimator is robust. As with Aït-Sahalia (1996a), we observe the diffusion function is globally an increasing function of the level of the interest rate between 0 and 14% and above 14%, the diffusion function flattens and then decreases.

The above comparisons give similar conclusions for the Euro data as for the Fed data. We suggest our optimal bandwidth is retrieved with $h_2 = \frac{1}{10} \times T^{-\frac{1}{5}}$ and that the linear mean-reverting specification for the drift is

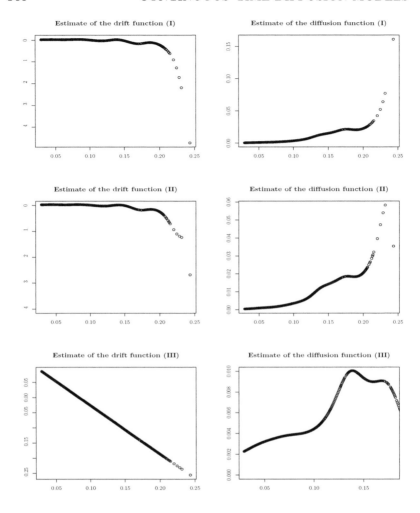

Figure 5.13 *Nonparametric drift and diffusion estimators for the Euro data with $h_a = 0.01347$.*

not applicable for high-frequency, shorter maturity proxies. Acceptance of the estimators from the first two methods suggests the CIR (1985) linear specification for the diffusion may be a good approximation for $r < 11\%$, while for larger interest rates, the quadratic specification of Brennan and Schwartz (1980) or the specification of Chan *et al.* (1992) may be applicable as the volatility increases at a faster rate than r_t. It may be appropriate to apply these models over the whole range of r.

5.4.5 Conclusions and discussion

Many different specification tests for the drift and diffusion functions of the common Itô diffusion process have opened up a new area of research. Some recently published papers empirically compare the plethora of proposed models.

In this study, we adopted a similar nonparametric approach as Aït-Sahalia (1996a) and Jiang and Knight (1997) to estimate the diffusion process. We used two popular short rates: the three–month Treasury Bill and the seven–day Eurodollar deposit rates. Based on our analysis, we suggested the use of the bandwidth $h = \frac{1}{10} \times T^{-\frac{1}{5}}$ for both sets of data. We then demonstrated how the bandwidth choice can have a dramatic effect on the drift and diffusion estimates. We rejected linearity in the drift despite theoretic economic justification, both empirically and more formally with the specification test L^*. Overall, we would suggest using the pair $(\widehat{\mu}_2(r), \widehat{\sigma}_3^2(r))$ for the two sets of data.

In summary, the nonparametric specification of the diffusion function of Method 1 is seen to differ significantly from the diffusion function estimates of Methods 2 and 3. We thus conclude that restrictions on the drift have a greater effect on the volatility than what is suggested in the literature. Our empirical comparisons suggest that $\widehat{\sigma}_1^2(\cdot)$ is misspecified primarily as a result of the assumptions imposed on it. That is, the assumption of a linear mean–reverting drift has a substantial effect on the final form of the diffusion. We suggest relaxing the drift assumption. The unrestricted drift estimates indicate that the fitting of a second–order polynomial may be more appropriate. It would be interesting to regenerate these estimators but with a quadratic restriction on the drift. The results could then be compared with the diffusion function estimates generated by Methods 2 and 3. Such an exercise is deferred to future research.

Extended research in this area should include a comparison of the nonparametric density, drift and diffusion estimators with those implied by some of the popular parametric models (see Aït-Sahalia 1999). In particular, to consider our nonparametrically estimated marginal density with, say, the Gamma density of CIR (1985) applied to the Eurodollar data, or to review how the unrestricted diffusion estimators actually compare to the diffusion specifications of CIR (1985), Brennan and Schwartz (1980) and Chan *et al.* (1992). Additionally, it would be useful to apply both the popular existing parametric models and our nonparametric estimates to price derivatives (e.g., bond options) in an attempt to determine the accuracy of the prices computed.

5.5 Technical notes

This section provides some necessary conditions for the establishment of the theorems as well as their proofs.

5.5.1 Assumptions

Assumption 5.1. (i) Assume that the process $\{r_t\}$ is strictly stationary and α-mixing with the mixing coefficient $\alpha(t) \leq C_\alpha \alpha^t$, where $0 < C_\alpha < \infty$ and $0 < \alpha < 1$ are constants.

(ii) The bandwidth parameter h satisfies that

$$\lim_{T\to\infty} h = 0 \quad \text{and} \quad \lim_{T\to\infty} Th^3 = \infty.$$

Assumption 5.2. (i) The marginal density function $\pi(r)$ is three times continuously differentiable in $r \in \mathbf{R}^+ = (0,\infty)$. In addition, $\pi(0) = 0$.

(ii) The drift and the diffusion functions $\mu(r)$ and $\sigma^2(r)$ are three times continuously differentiable in $r \in \mathbf{R}^+ = (0,\infty)$, and $\sigma(r) > 0$ on $\mathbf{R}^+$.

(iii) The integral of $\bar{\mu}(v) = \frac{1}{\sigma^2(v)} \exp\left(-\int_v^{\bar{v}} 2\frac{\mu(x)}{\sigma^2(x)} dx\right)$ converges at both boundaries of $\mathbf{R}^+$, where $\bar{v}$ is fixed in $\mathbf{R}^+$.

(iv) The integral of $s(v) = \exp\left(\int_v^{\bar{v}} 2\frac{\mu(x)}{\sigma^2(x)} dx\right)$ diverges at both boundaries of $\mathbf{R}^+$.

Assumption 5.3. (i) The second derivative of $\pi(r)$, $\pi''(r)$, is square integrable over $\mathbf{R}^+$.

(ii) The following functions are integrable over $\mathbf{R}^+$ for $i = 1, 2$:

$$(r\pi'(r))^2 \quad \text{and} \quad \left(\int_0^r \pi''(u)du\right)^2.$$

(iii) The following functions are integrable over $\mathbf{R}^+$:

$$\mu^4(r)\pi(r), \quad \sigma^4(r)\pi(r), \quad \left(\frac{d^2}{dr^2}[\mu(r)\pi(r)]\right)^2,$$

$$\left(\frac{d^i}{dr^i}[\mu^2(r)\pi(r)]\right)^2, \quad \left(\frac{d^i}{dr^i}[\sigma^2(r)\pi(r)]\right)^2$$

for $i = 1, 2, 3$, and

$$\left(\int_0^r \frac{d^2}{dx^2}[\mu(x)\pi(x)]dx\right)^2.$$

Assumption 5.4. (i) For $\psi(r) = \sigma^2(r)$ or $\sigma^4(r)$, $\psi(r)$ satisfies the Lipschitz type condition: $|\psi(r+v) - \psi(r)| \leq \Psi(r)|v|$ for $v \in S$ (any compact subset of $\boldsymbol{R}^1$), where $\Psi(r)$ is a measurable function such that $E\left[\Psi^2(r)\right] < \infty$.

(ii) Assume that the set H_T has the structure of (5.48) with

$$c_{\max}(\log\log T)^{-1} = h_{\max} > h_{\min} \geq T^{-\gamma}$$

for some constant γ such that $0 < \gamma < \frac{1}{3}$.

Assumptions 5.1–5.4 are natural in this kind of problem. Assumption 5.1(i) assumes the α–mixing condition, which is weaker than the β–mixing condition. Assumption 5.1(ii) ensures that the theoretically optimum value of $h = c \cdot T^{-1/5}$ can be used. Assumption 5.2 is equivalent to Assumption A1 of Aït-Sahalia (1996a), requiring the existence and uniqueness of a strong solution to model (5.1). Assumption 5.3 basically requires that all the integrals involved in Theorems 5.1–5.6 do exist. Assumption 5.4 is used only for the establishment and proof of Theorems 5.6 and 5.7. Similar conditions have been assumed in Assumptions 2 and 6 of Horowitz and Spokoiny (2001).

5.5.2 Proof of Equation (5.11)

Keeping in mind that $K\left(\frac{x}{h}\right) = \frac{1}{\sqrt{2\pi}}\exp\{-\frac{x^2}{2h^2}\}$ and

$$\widehat{\pi}(r) = \frac{1}{Th}\sum_{t=1}^{T}K\left(\frac{r-X_t}{h}\right),$$

we now derive our estimator, $\widehat{\sigma}_1^2(\cdot)$. Recall that Equation (5.9) implies

$$\widehat{\sigma}_1^2(r) = \frac{2}{\widehat{\pi}(r)}\int_0^r \mu(u;\widehat{\theta})\widehat{\pi}(u)du.$$

Now, evaluating the integral on the right of this identity,

$$\int_0^r \mu(u;\widehat{\theta})\widehat{\pi}(u)du = \int_0^r \widehat{\beta}(\widehat{\alpha} - u)\frac{1}{Th}\sum_{t=1}^{T}K\left(\frac{u-X_t}{h}\right)du$$

$$= \frac{\widehat{\beta}}{Th}\sum_{t=1}^{T}\int_0^r (\widehat{\alpha}-u)K\left(\frac{u-X_t}{h}\right)du$$

$$= \frac{\widehat{\beta}}{Th}\sum_{t=1}^{T}\int_0^r (\widehat{\alpha}-X_t+X_t-u)K\left(\frac{u-X_t}{h}\right)du$$

$$= \frac{\widehat{\beta}}{Th} \sum_{t=1}^{T} \left(\int_0^r (\widehat{\alpha} - X_t) K\left(\frac{u - X_t}{h}\right) du \right)$$

$$+ \frac{\widehat{\beta}}{Th} \sum_{t=1}^{T} \left(\int_0^r (X_t - u) K\left(\frac{u - X_t}{h}\right) du \right)$$

$$= \frac{\widehat{\beta}}{Th} \sum_{t=1}^{T} \left((\widehat{\alpha} - X_t) \int_0^r K\left(\frac{u - X_t}{h}\right) du \right)$$

$$- \frac{\widehat{\beta}}{Th} \sum_{t=1}^{T} \left(\int_0^r [u - X_t] K\left(\frac{u - X_t}{h}\right) du \right)$$

$$= \frac{\widehat{\beta}}{Th} \sum_{t=1}^{T} (\widehat{\alpha} - X_t) h \int_{-\frac{X_t}{h}}^{\frac{r - X_t}{h}} K(v) dv$$

$$- \frac{\widehat{\beta}}{Th} \sum_{t=1}^{T} h^2 \int_{-\frac{X_t}{h}}^{\frac{r - X_t}{h}} v K(v) dv$$

$$= \frac{\widehat{\beta}}{T} \sum_{t=1}^{T} \left((\widehat{\alpha} - X_t) \int_{-\frac{X_t}{h}}^{\frac{r - X_t}{h}} \frac{\exp\left\{-\frac{v^2}{2}\right\}}{\sqrt{2\pi}} dv \right)$$

$$- \frac{\widehat{\beta}}{T} \sum_{t=1}^{T} h \int_{-\frac{X_t}{h}}^{\frac{r - X_t}{h}} \frac{v}{\sqrt{2\pi}} \exp\left\{-\frac{v^2}{2}\right\} dv$$

$$= \frac{\widehat{\beta}}{T} \sum_{t=1}^{T} (\widehat{\alpha} - X_t) \left[\Phi\left(\frac{r - X_t}{h}\right) - \Phi\left(-\frac{X_t}{h}\right) \right]$$

$$- \frac{\widehat{\beta}}{T} \sum_{t=1}^{T} h I_{1t},$$

where $\Phi(x) = \frac{1}{\sqrt{2\pi}} \int_{-\infty}^x \exp\{-\frac{u^2}{2}\} du$ and

$$I_{1t} = \frac{1}{\sqrt{2\pi}} \int_{-\frac{X_t}{h}}^{\frac{r - X_t}{h}} v \exp\left\{-\frac{v^2}{2}\right\} dv.$$

Observe that

$$I_{1t} = \int_{-\frac{X_t}{h}}^{\frac{r - X_t}{h}} \frac{v}{\sqrt{2\pi}} \exp\left\{-\frac{v^2}{2}\right\} dv = -\frac{1}{\sqrt{2\pi}} \int_{-\frac{X_t}{h}}^{\frac{r - X_t}{h}} d\left(\exp\left\{-\frac{v^2}{2}\right\}\right)$$

$$= -\frac{1}{\sqrt{2\pi}} \left[\exp\left\{-\frac{(r - X_t)^2}{2h^2}\right\} - \exp\left\{-\frac{X_t^2}{2h^2}\right\} \right].$$

Therefore,

$$
\int_0^r \mu(u;\widehat{\theta})\widehat{\pi}(u)du \;=\; \frac{\widehat{\beta}}{T}\sum_{t=1}^{T}\left\{\left[\widehat{\alpha}-X_t\right]\left[\Phi\left(\frac{r-X_t}{h}\right)-\Phi\left(-\frac{X_t}{h}\right)\right]\right.
$$

$$
+\;\frac{h}{\sqrt{2\pi}}\left[\exp\left\{-\frac{(r-X_t)^2}{2h^2}\right\}-\exp\left\{-\frac{X_t^2}{2h^2}\right\}\right]\right\}
$$

$$
=\;\frac{\widehat{\beta}}{T}\sum_{t=1}^{T}\left[\widehat{\alpha}-X_t\right]\left[\Phi\left(\frac{r-X_t}{h}\right)-\Phi\left(-\frac{X_t}{h}\right)\right]
$$

$$
+\;\frac{\widehat{\beta}\,h}{T\sqrt{2\pi}}\sum_{t=1}^{T}\left[\exp\left\{-\frac{(r-X_t)^2}{2h^2}\right\}-\exp\left\{-\frac{X_t^2}{2h^2}\right\}\right].
$$

This gives

$$
\widehat{\sigma}_1^2(r) \;=\; \frac{2\widehat{\beta}}{T\widehat{\pi}(r)}\cdot\sum_{t=1}^{T}\left[\widehat{\alpha}-X_t\right]\left[\Phi\left(\frac{r-X_t}{h}\right)-\Phi\left(\frac{-X_t}{h}\right)\right]
$$

$$
+\;\frac{2\widehat{\beta}}{T\widehat{\pi}(r)}\cdot\frac{h}{\sqrt{2\pi}}\sum_{t=1}^{T}\left[\exp\left(-\frac{(r-X_t)^2}{2h^2}\right)-\exp\left(-\frac{X_t^2}{2h^2}\right)\right].
$$

5.5.3 Proofs of Theorems 5.2–5.5

As there are some similarities among the proofs of Theorems 5.2–5.5, we provide only an outline for the proof of Theorems 5.4 and 5.5 in some detail. However, the details of the other proofs are available upon request.

Recall that $Y_t = \frac{r_{(t+1)\Delta}-r_{t\Delta}}{\Delta}$ and observe that $Y_t = \mu(X_t) + \sigma(X_t)e_t$, where $X_t = r_{t\Delta}$ and $e_t = \frac{B_{(t+1)\Delta}-B_{t\Delta}}{\Delta}$. We now have

$$
\widehat{m}_3(r) \;=\; \widehat{\mu}_3(r)\widehat{\pi}(r) = \frac{\Delta}{2Th^2}\sum_{t=1}^{T-1}\frac{(X_t-r)}{h}K\left(-\frac{(X_t-r)}{h}\right)Y_t^2
$$

$$
=\;\frac{\Delta}{2Th^2}\sum_{t=1}^{T-1}\frac{(X_t-r)}{h}K\left(-\frac{(X_t-r)}{h}\right)
$$

$$
\times\;\left[2\mu(X_t)\sigma(X_t)+\mu^2(X_t)+\sigma^2(X_t)e_t^2\right]
$$

$$
=\;\frac{\Delta}{Th^2}\sum_{t=1}^{T-1}\frac{(X_t-r)}{h}K\left(-\frac{(X_t-r)}{h}\right)\mu(X_t)\sigma(X_t)e_t
$$

$$
+\;\frac{\Delta}{2Th^2}\sum_{t=1}^{T-1}\frac{(X_t-r)}{h}K\left(-\frac{(X_t-r)}{h}\right)\left[\mu^2(X_t)+\sigma^2(X_t)e_t^2\right].
$$

Thus, a Taylor expansion implies that as $h \to 0$

$$
\begin{aligned}
E\left[\widehat{m}_3(r)\right] &= \frac{-\Delta}{2h^2} \int \frac{r-s}{h} K\left(\frac{r-s}{h}\right) [\mu^2(s) + \sigma^2(s)\sigma_0^2]\pi(s)ds \\
&= \frac{-\Delta}{2h} \int xK(x) \left[\mu^2(r-xh) + \sigma^2(r-hx)\sigma_0^2\right] \pi(r-hx)dx \\
&= \frac{-\Delta}{2h} \left(-hp'(r) + \frac{h^3}{6} \int x^4 K(x)p^{(3)}(\xi)dx\right) \\
&= \frac{\Delta}{2}p'(r) + \frac{\Delta}{4}p^{(3)}(r)h^2 + o(h^2),
\end{aligned}
\tag{5.50}
$$

where $\sigma_0^2 = E[e_t^2] = \Delta^{-1}$, $p(r) = \mu^2(r) + \sigma^2(r)\sigma_0^2$, and ξ lies between $r - hx$ and r.

This implies

$$
\begin{aligned}
E\left[\widehat{m}_3(r)\right] &= m(r) + \frac{\Delta}{2}\frac{d}{dr}[\mu^2(r)\pi(r)] \\
&+ \frac{\Delta}{4}p^{(3)}(r)h^2 + o(h^2).
\end{aligned}
\tag{5.51}
$$

Let $Z_t = \left(\frac{X_t-r}{h}\right) K\left(-\frac{X_t-r}{h}\right)\mu(X_t)\sigma(X_t)e_t$ and

$$
W_t = \left(\frac{X_t - r}{h}\right) K\left(-\frac{X_t - r}{h}\right) \left[\mu^2(X_t) + \sigma^2(X_t)e_t^2\right].
$$

Observe that

$$
\begin{aligned}
\widehat{m}_3(r) - E\left[\widehat{m}_3(r)\right] &= \frac{\Delta}{Th^2}\sum_{t=1}^{T-1} Z_t + \frac{\Delta}{2Th^2}\sum_{t=1}^{T-1}(W_t - E[W_t]) \\
&\equiv I_{1T} + I_{2T},
\end{aligned}
\tag{5.52}
$$

where the symbol " $\equiv$ " denotes that the terms of the left–hand side are correspondingly identical to those of the right–hand side.

Analogously to (5.50), we obtain that as $h \to 0$

$$
\begin{aligned}
E[I_{1T}^2] &= \frac{\Delta^2}{T^2h^4}(1 + o(1))\sum_{t=1}^{T-1} E\left[Z_t^2\right] + \sum_{t=1}^{T-1}\sum_{s=1,\neq t}^{T-1} E\left[Z_s Z_t\right] \\
&= \frac{\Delta^2\sigma_0^2}{Th^4} \int \left(\frac{s-r}{h}\right)^2 K^2\left(\frac{r-s}{h}\right)\mu^2(s)\sigma^2(s)\pi(s)ds \\
&\times (1 + o(1)) \\
&= \frac{\Delta^2\sigma_0^2}{Th^3}q(r) \int x^2 K^2(x)dx + \frac{\Delta^2\sigma_0^2}{Th}q''(r) \int x^4 K^2(x)dx \\
&\times (1 + o(1)),
\end{aligned}
\tag{5.53}
$$

where $q(r) = \mu^2(r)\sigma^2(r)\pi(r)$. Similarly, we can show that as $h \to 0$

$$
\begin{aligned}
E[I_{2T}^2] &= \frac{\Delta^2}{4Th^3}(1 + o(1))\left[\mu^4(r) + 3\sigma^4(r)\sigma_0^4 + 2\mu^2(r)\sigma^2(r)\sigma_0^2\right] \\
&\quad \times \int x^2 K^2(x)dx.
\end{aligned}
\tag{5.54}
$$

Equations (5.50)–(5.54) then imply Theorem 5.4. Observe that

$$
\begin{aligned}
E\left[\widehat{V}_3(r)\right] &= \frac{2}{h}E\left[\mu(X_t)\int_0^r K\left(\frac{u - X_t}{h}\right)du\right] \\
&= \frac{2}{h}\int_o^r\left[\int K\left(\frac{u - v}{h}\right)\mu(v)\pi(v)dv\right]du \\
&= 2\int_0^r \mu(u)\pi(u)du + h^2\int_0^r \frac{d^2}{du^2}[\mu(u)\pi(u)]du \\
&= V(r) + h^2\int_0^r \frac{d^2}{du^2}[\mu(u)\pi(u)]du + o(h^2)
\end{aligned}
\tag{5.55}
$$

using a Taylor expansion.

Let $\Psi(X_t) = \Phi\left(\frac{r - X_t}{h}\right) - \Phi\left(\frac{-X_t}{h}\right)$. Similarly to (5.53) and (5.54), we obtain that for sufficiently large T

$$
\begin{aligned}
E\left(\widehat{V}_3(r) - E\left[\widehat{V}_3(r)\right]\right)^2 &= E\left\{\frac{2}{T}\sum_{t=1}^{T-1}(Y_t\Psi(X_t) - E[Y_t\Psi(X_t)])\right\}^2 \\
&= \frac{4}{T}\int\left[\mu^2(s) + \Delta^{-1}\sigma^2(s)\right]\Psi^2(s)\pi(s)ds \\
&\quad + o\left(\frac{1}{T}\right).
\end{aligned}
\tag{5.56}
$$

Theorem 5.5 then follows from (5.55), (5.56) and

$$
E\left(\widehat{V}_3(r) - V(r)\right)^2 = E\left(\widehat{V}_3(r) - E\left[\widehat{V}_3(r)\right]\right)^2 + \left(E\left[\widehat{V}_3(r)\right] - V(r)\right)^2.
$$

5.5.4 Technical lemmas

This section lists the key lemmas but provides only an outline of the proof of each lemma, as the detailed proofs of the lemmas are extremely technical and therefore omitted here. However, they are available from the authors upon request. The proof of Lemma 5.1 below follows similarly from that of Theorem 3.1(i) of Li (1999). The proofs of Lemmas 5.2–5.6 below follow similarly from those of Lemmas 8, 10, 12 and 13 of Horowitz and Spokoiny (HS) (2001), respectively. Note that Assumption

1 of HS (2001) holds automatically since $\mu(r, \theta)$ of this paper is a linear function of θ. Also note that Assumption 2 of HS (2001) holds immediately as $\widehat{\theta}$ is the least–squares estimator. In addition, we need not assume the compactness condition on K as in Assumption 4 of HS (2001). This is because there is no denominator involved in the numerator part of the form of $L_{53}(h)$ as can be seen from the proof of Lemma 5.1 below.

Lemma 5.1. *Suppose that Assumptions 5.1(i), 5.2 and 5.4(ii) hold. Then under* $\mathcal{H}_{053}$

$$L_{53}(h) \to_D N(0, 1)$$

as $T \to \infty$ *and for every given* $h \in H_T$.

PROOF: Let $\epsilon_t = \sigma(X_t)e_t$. Recall $X_t = r_{t\Delta}$ and observe that for any $\theta \in \Theta$

$$
\begin{aligned}
\widehat{\epsilon}_t &= Y_t - \mu(X_t; \widehat{\theta}) = \epsilon_t + \mu(X_t; \theta) - \mu(X_t; \widehat{\theta}) \\
&= \epsilon_t + \mu(X_t; \theta) - \mu(X_t; \theta_0) + \mu(X_t; \theta_0) - \mu(X_t; \widehat{\theta}) \\
&\equiv \epsilon_t + \lambda_t(\theta) + \delta_t, \quad\quad\quad (5.57)
\end{aligned}
$$

where $\lambda_t(\theta) = \mu(X_t; \theta) - \mu(X_t; \theta_0)$ and $\delta_t = \mu(X_t; \theta_0) - \mu(X_t; \widehat{\theta})$.

Let $S_{53}^2 = 2\sum_{s=1}^{T}\sum_{t=1}^{T}\widehat{\epsilon}_s^2 \, K^2\left(\frac{X_s - X_t}{h}\right) \widehat{\epsilon}_t^2$. It then follows from the definition of $L_{53}(h)$ that for sufficiently large T,

$$
\begin{aligned}
L_{53}(h) &= \frac{\sum_{s=1}^{T}\sum_{t=1, t\neq s}^{T}\widehat{\epsilon}_s \, K\left(\frac{X_s - X_t}{h}\right) \widehat{\epsilon}_t}{\sqrt{2\sum_{s=1}^{T}\widehat{\epsilon}_s^2 \sum_{t=1}^{T} K^2\left(\frac{X_s - X_t}{h}\right) \widehat{\epsilon}_t^2}} \\[2mm]
&= \frac{1}{S_{53}}\sum_{s=1}^{T}\sum_{t=1, t\neq s}^{T}\epsilon_s \, K\left(\frac{X_s - X_t}{h}\right)\epsilon_t \\[2mm]
&+ \frac{1}{S_{53}}\sum_{s=1}^{T}\sum_{t=1, t\neq s}^{T}\lambda_s(\theta) \, K\left(\frac{X_s - X_t}{h}\right)\lambda_t(\theta) \\[2mm]
&+ \frac{1}{S_{53}}\sum_{s=1}^{T}\sum_{t=1, t\neq s}^{T}\delta_s \, K\left(\frac{X_s - X_t}{h}\right)\delta_t + o_P(L_T(h)) \\[2mm]
&= \frac{1}{S_{53}}\sum_{s=1}^{T}\sum_{t=1, t\neq s}^{T}\epsilon_s K\left(\frac{X_s - X_t}{h}\right)\epsilon_t \\[2mm]
&+ \frac{1}{S_{53}}\sum_{s=1}^{T}\sum_{t=1, t\neq s}^{T}\lambda_s(\theta) K\left(\frac{X_s - X_t}{h}\right)\lambda_t(\theta) \\[2mm]
&+ o_P(L_{53}(h)) \quad\quad\quad (5.58)
\end{aligned}
$$

using the fact that $\widehat{\theta}$ is a $\sqrt{T}$–consistent estimator of θ_0.

Similarly, we may have for sufficiently large T,

$$
\begin{aligned}
S_{53}^2 &= 2\sum_{s=1}^{T}\sum_{t=1}^{T}\hat{\epsilon}_s^2\, K^2\left(\frac{X_s - X_t}{h}\right)\hat{\epsilon}_t^2 \\
&= 2\sum_{s=1}^{T}\sum_{t=1}^{T}\epsilon_s^2\, K^2\left(\frac{X_s - X_t}{h}\right)\epsilon_t^2 \\
&+ 2\sum_{s=1}^{T}\sum_{t=1}^{T}\lambda_s^2(\theta)\, K^2\left(\frac{X_s - X_t}{h}\right)\lambda_t^2(\theta) \\
&+ 2\sum_{s=1}^{T}\sum_{t=1}^{T}\delta_s^2\, K^2\left(\frac{X_s - X_t}{h}\right)\delta_t^2 + o_P(S_T^2) \\
&= 2\sum_{s=1}^{T}\sum_{t=1}^{T}\epsilon_s^2\, K^2\left(\frac{X_s - X_t}{h}\right)\epsilon_t^2 \\
&+ 2\sum_{s=1}^{T}\sum_{t=1}^{T}\lambda_s^2(\theta)\, K^2\left(\frac{X_s - X_t}{h}\right)\lambda_t^2(\theta) + o_P(S_T^2). \quad (5.59)
\end{aligned}
$$

Analogously to the proof of Theorem 3.1(i) of Li (1999), we may show that under $\mathcal{H}_{053}$

$$
\lim_{T\to\infty}\frac{S_{53}^2}{\sigma_h^2} = 1 \quad \text{in probability,} \quad (5.60)
$$

where

$$
\begin{aligned}
\sigma_h^2 &= E\left[\sum_{s=1}^{T}\sum_{t=1}^{T}\epsilon_s^2\, K^2\left(\frac{X_s - X_t}{h}\right)\epsilon_t^2\right] \\
&= 2\sigma_0^4\, T(T-1)h\,(1 + o(1)) \\
&\times \int_{-\infty}^{\infty} K^2(u)du \int \sigma^4(v)\pi^2(v)dv, \quad (5.61)
\end{aligned}
$$

in which $\sigma_0^2 = E[e_t^2]$.

Since $\lambda(\theta_0) = 0$ under $\mathcal{H}_{053}$, Equations (5.58) and (5.59) imply that under $\mathcal{H}_{053}$

$$
\begin{aligned}
L_{53}(h) &= \frac{\sum_{s=1}^{T}\sum_{t=1,t\neq s}^{T}\epsilon_s\, K\left(\frac{X_s - X_t}{h}\right)\epsilon_t}{\sqrt{2\sum_{s=1}^{T}\sum_{t=1}^{T}\epsilon_s^2\, K^2\left(\frac{X_s - X_t}{h}\right)\epsilon_t^2}} + o_P(1) \quad (5.62) \\
&= \frac{\sum_{s=1}^{T}\sum_{t=1,t\neq s}^{T}u_s\,\sigma(X_s)K\left(\frac{X_s - X_t}{h}\right)\sigma(X_t)\,u_t}{\sqrt{2\sum_{s=1}^{T}\sum_{t=1}^{T}u_s^2\,\sigma^2(X_s)K^2\left(\frac{X_s - X_t}{h}\right)\sigma^2(X_t)\,u_t^2}} + o_P(1)
\end{aligned}
$$

for sufficiently large T, where $u_t = \sqrt{\Delta}\, e_t \sim N(0,1)$ is independent of Δ.

Let $p_{st} = \sigma(X_s) K\left(\frac{X_s - X_t}{h}\right) \sigma(X_t)$ and $\phi_{st} = u_s\, p_{st}\, u_t$. Equations (5.57)–(5.62) imply that under $\mathcal{H}_{053}$

$$L_{53}(h) = \frac{1}{S_{53}} \sum_{t=1}^{T} \sum_{s=1,\neq t}^{T} \hat{\epsilon}_s\, p_{st}\, \hat{\epsilon}_t = \frac{1}{\sigma_{1h}} \sum_{t=1}^{T} \sum_{s=1,\neq t}^{T} \phi_{st} + o_P(1) \quad (5.63)$$

for sufficiently large T, where

$$\sigma_{1h}^2 = \frac{\sigma_h^2}{\sigma_0^4} = 2T(T-1)h\,(1+o(1)) \int_{-\infty}^{\infty} K^2(u)du \int \sigma^4(v)\pi^2(v)dv$$

is independent of Δ. Note that Theorem A.1 of the appendix is applicable to such ϕ_{st}. The proof of our Lemma 5.1 then follows similarly from that of Theorem A.1 of the appendix. The detail is similar to the proof of Theorem 2.1 of Gao and King (2004).

Before establishing some other lemmas, we need to introduce some additional symbols and notation. Define

$$N_{0T}(h) = \sum_{s=1}^{T} \sum_{t=1,t\neq s}^{T} \epsilon_s\, K\left(\frac{X_s - X_t}{h}\right) \epsilon_t,$$

$$Q_T(\theta) = \sum_{s=1}^{T} \sum_{t=1,t\neq s}^{T} \lambda_s(\theta)\, K\left(\frac{X_s - X_t}{h}\right) \lambda_t(\theta),$$

$$D_{0T}(h,\theta) = 2\sum_{s=1}^{T} \sum_{t=1}^{T} \epsilon_s^2\, K^2\left(\frac{X_s - X_t}{h}\right) \epsilon_t^2$$

$$+ 2\sum_{s=1}^{T} \sum_{t=1}^{T} \lambda_s^2(\theta)\, K^2\left(\frac{X_s - X_t}{h}\right) \lambda_t^2(\theta).$$

Let $N_{0T}^*(h)$ be the version of $N_{0T}(h)$ with $\{e_t\}$ replaced by $\{e_t^*\}$. Define

$$L_0(h) = \frac{N_{0T}(h)}{\sigma_h}, \qquad L_0^*(h) = \frac{N_{0T}^*(h)}{\sigma_h}, \qquad (5.64)$$

$$\widehat{L}_0(h) = \widehat{L}_0(h,\theta) = \frac{N_{0T}(h) + Q_T(\theta)}{\sqrt{D_{0T}(h,\theta)}}, \qquad (5.65)$$

where $\theta \in \Theta$.

In addition, as proposed in the simulation procedure below (5.48), let $\widehat{L}^*(h)$ be the version of $L_{53}(h)$ with Y_t and $\widehat{\theta}$ replaced by Y_t^* and $\widehat{\theta}^*$ on the right–hand side of (5.47).

Lemma 5.2. *Suppose that Assumptions 5.1(i), 5.2 and 5.4 hold. Then*

$$L^* = \max_{h \in H_T} L_{53}(h) = \max_{h \in H_T} \widehat{L}_0(h, \theta) + o_p(1), \qquad (5.66)$$

$$\widehat{L}^* = \max_{h \in H_T} \widehat{L}^*(h) = \max_{h \in H_T} L_0^*(h) + o_p(1). \qquad (5.67)$$

PROOF: The proof of (5.66) and (5.67) follows from (5.57)–(5.61). The detail is similar to that of Lemma A.2 of Arapis and Gao (2006).

Lemma 5.3. *Let Assumptions 5.1–5.2 and 5.4 hold. Then the asymptotic distributions of $\max_{h \in H_T} L_0(h)$ and $\max_{h \in H_T} L_0^*(h)$ are identical under $\mathcal{H}_{053}$.*

PROOF: The proof follows easily from Lemma 5.2 and the fact that both $\{e_t\}$ and $\{e_t^*\}$ are mutually independent and identically distributed as $N\left(0, \Delta^{-1}\right)$.

Lemma 5.4. *Suppose that Assumptions 5.1(i), 5.2 and 5.4(ii) hold. Then for any $x \geq 0$, $h \in H_T$ and all sufficiently large T*

$$P\left(L_0^*(h) > x\right) \leq \exp\left(-\frac{x^2}{4}\right).$$

PROOF: The proof follows from the fact that $L_0^*(h)$ is asymptotically normal. The detailed proof is similar to that of Lemma A.4 of Arapis and Gao (2006).

For $0 < \alpha < 1$, define $\tilde{l}_\alpha$ to be the $1 - \alpha$ quantile of $\max_{h \in H_T} L_0^*(h)$.

Lemma 5.5. *Suppose that Assumptions 5.1(i), 5.2 and 5.4(ii) hold. Then for large enough T*

$$\tilde{l}_\alpha \leq 2\sqrt{\log(J_T) - \log(\alpha)}.$$

PROOF: The proof is trivial.

Lemma 5.6. *Suppose that Assumptions 5.1(i), 5.2 and 5.4(ii) hold. Suppose that*

$$\lim_{T \to \infty} P\left(\frac{Q_T(\theta_1)}{\sigma_h} \geq 2\tilde{l}_\alpha^*\right) = 1 \qquad (5.68)$$

for some $\theta_1 \in \Theta$ and $h \in H_T$, where

$$\tilde{l}_\alpha^* = \max\left(\tilde{l}_\alpha, \sqrt{2\log(J_T) + \sqrt{2\log(J_T)}}\right). \qquad (5.69)$$

Then

$$\lim_{T \to \infty} P(L^* > l_\alpha^*) = 1.$$

PROOF: The proof is similar to that of Lemma A.6 of Arapis and Gao (2006), in view of the proof of Lemma 13 of Horowitz and Spokoiny (2001).

5.5.5 Proofs of Theorems 5.6 and 5.7

Proof of Theorem 5.6: The proof follows from Lemmas 5.2 and 5.3.

Proof of Theorem 5.7: Let $\lambda(\theta_1) = (\lambda_1(\theta_1), \cdots, \lambda_T(\theta_1))^\tau$. In view of the definition of $Q_T(\theta)$, we have that for sufficiently large T

$$
\begin{aligned}
Q_T(\theta_1) &= (1 + o_P(1))\ Th\ \lambda(\theta_1)^\tau \lambda(\theta_1) \\
&= (1 + o_P(1))\ C_1\ T^2 h\ \geq C_2\ Th^{1/2} \sqrt{\log\log(T)} \quad (5.70)
\end{aligned}
$$

hold in probability, where $C_1 > 0$ and $C_2 > 0$ are constants. Equation (5.70), together with Assumption 5.4(ii) and (5.69), implies (5.68). This therefore completes the proof of Theorem 5.7.

5.6 Bibliographical notes

On nonparametric model specification of continuous–time diffusion models, recent developments include Aït-Sahalia (1996b), Jiang and Knight (1997), Stanton (1997), Jiang (1998), Pritsker (1998), Corradi and White (1999), Chapman and Pearson (2000), Gao (2000), Engle (2001), Bandi and Phillips (2003), Dette and Von Lieres und Wilkau (2003), Fan and Zhang (2003), Durham (2004), Gao and King (2004), Chen and Gao (2005), Corradi and Swanson (2005), Fan (2005), Hong and Li (2005), Jones (2005), and Li, Pearson and Poteshman (2005).

Other closely related studies about nonparametric and semiparametric estimation and testing in continuous-time diffusion models include Aït-Sahalia and Lo (1998, 2000), Aït-Sahalia (1999), Sundaresan (2001), Cai and Hong (2003), Fan and Zhang (2003), Fan et al. (2003), Kristensen (2004), Fan (2005), and others.

CHAPTER 6

Long–Range Dependent Time Series

6.1 Introductory results

There is a long history of studying long-range dependent models for strictly stationary time series. Hurst (1951), Mandelbrot and Van Ness (1968), Granger and Joyeux (1980), and Geweke and Porter–Hudak (1983) were among the first to study time series models with long-range dependence through using the spectral density approach. Attention has recently been given to two single-parameter models in which the spectral density function is proportional to $\omega^{-\gamma}$, $1 < \gamma < 2$, for ω near zero, and the asymptotic decay of the autocorrelation function is proportional to $\tau^{\gamma-1}$. Because the spectral density function is unbounded at $\omega = 0$– equivalently, the autocorrelation function is not summable, these are long–range dependent (LRD) (long memory; strong dependent) models. A recently published survey of long-range dependence literature up to about 1994 is Beran (1994). See also Robinson (1994), Baillie and King (1996), Anh and Heyde (1999) and Robinson (2003) for recent developments of long-range dependence in econometrics and statistics.

We now state two definitions and a theorem, which are the same as Definitions 2.1 and 2.2 as well as Theorem 2.1 of Beran (1994).

If $\{Z_t\}$ is a zero mean, real-valued, and stationary process, it has an autocovariance function

$$r(\tau) = E[Z_t Z_{t+\tau}]$$

and the spectral representation of the form (see Theorem 4.3.2 of Brockwell and Davis 1990)

$$r(\tau) = \int_{-\pi}^{\pi} e^{i\omega\tau} dF(\omega), \qquad (6.1)$$

where $F(\omega)$ is nondecreasing and bounded. If $F(\omega)$ is absolutely continuous, its density $f(\omega)$ is called the spectrum of Z_t. There is an inversion formula

$$f(\omega) = \frac{1}{2\pi} \sum_{n=-\infty}^{\infty} r(n) e^{-2in\omega} \tag{6.2}$$

provided that $r(\tau)$ is absolutely summable. For this case, $\{Z_t\}$ is called a stationary process with *short–range dependence*.

We now state the following definitions for the case where $r(\tau)$ is not absolutely summable. Let $\rho(\tau) = \frac{r(\tau)}{r(0)}$.

Definition 6.1. Let $\{Z_t\}$ be a stationary process for which the following holds. There exists a real number $\alpha \in (0,1)$ and a constant $C_\rho > 0$ such that

$$\lim_{\tau \to \infty} \frac{\rho(\tau)}{C_\rho \tau^{-\alpha}} = 1. \tag{6.3}$$

Then $\{Z_t\}$ is called a stationary process with *long–range dependence*. Equation (6.3) can be written as

$$\rho(\tau) \approx C_\rho \frac{1}{\tau^\alpha} \text{ as } \tau \to \infty,$$

where the symbol "$\approx$" indicates that the ratio tends to one as $\tau \to \infty$.

Knowing the autocorrelations is equivalent to knowing $f(\omega)$. Therefore, long-range dependence can also be defined by imposing a condition on the spectral density.

Definition 6.2. Let $\{Z_t\}$ be a stationary process for which the following holds. There exists a real number $\beta \in (0,1)$ and a constant $C_f > 0$ such that

$$\lim_{\lambda \to 0} \frac{f(\omega)}{C_f |\omega|^{-\beta}} = 1. \tag{6.4}$$

Then $\{Z_t\}$ is called a stationary process with long-range dependence. Equation (6.4) can be written as

$$f(\omega) \approx C_f \frac{1}{|\omega|^\beta} \text{ as } \omega \to 0.$$

These two definitions are equivalent in the following sense.

Theorem 6.1. (i) *Suppose (6.3) holds with* $0 < \alpha = 2 - 2H < 1$. *Then the spectral density* f *exists and*

$$\lim_{\omega \to 0} \frac{f(\omega)}{C_f |\omega|^{1-2H}} = 1,$$

where

$$C_f = \sigma^2 \pi^{-1} C_\rho \Gamma(2H - 1) \sin(\pi - \pi H) \text{ with } \sigma^2 = \text{var}(Z_t).$$

(ii) *Suppose (6.4) holds with* $0 < \beta = 2H - 1 < 1$. *Then*

$$\lim_{\tau \to \infty} \frac{\rho(\tau)}{C_\rho \tau^{2H-2}} = 1,$$

where

$$C_\rho = 2C_f \Gamma(2 - 2H) \sin(\pi H - 0.5\pi)\sigma^{-2}.$$

The proof of Theorem 6.1 is straightforward. It is important to note that the definition of long-range dependence by (6.3) (or Equation (6.4)) is an asymptotic definition and that the relation $\alpha + \beta = 1$ is always true. The parameter H is called a self–similarity parameter.

6.2 Gaussian semiparametric estimation

Let $\{Z_t\}$ be a stationary long–range dependent process with zero mean. Denote by $r(\tau)$ the lag–τ autocovariance of $\{Z_t\}$ and by $f_\theta(\omega)$ the spectral density of $\{Z_t\}$ such that

$$r(\tau) = E[Z_t Z_{t+\tau}] = \int_{-\pi}^{\pi} \cos(\tau\omega) f_\theta(\omega) \, d\omega.$$

It is assumed that

$$f_\theta(\omega) \sim G \, \omega^{1-2H} \quad \text{as} \quad \omega \to 0+, \tag{6.5}$$

where $G \in (0, \infty)$, $H \in \left(\frac{1}{2}, 1\right)$ and $\theta = (G, H) \in \Theta = (0, \infty) \times \left(\frac{1}{2}, 1\right)$. The parameter H is sometimes called the self-similarity parameter. The estimation of G and H has been popular in recent years. There are a number of different approaches to the estimation of G and H. See, for example, Geweke and Porter–Hudak (1983), Fox and Taqqu (1986), Dahlhaus (1989), Heyde and Gay (1993), and Robinson (1995a). This section suggests using the Gaussian semiparametric estimation method proposed by Robinson (1995b) as this estimation procedure is more efficient and its asymptotic properties can be established in a broader context.

The periodogram of $\{Z_t\}$ is defined at the Fourier frequencies $\omega_j = \frac{2\pi j}{T} \in (-\pi, \pi]$, by

$$I_T(\omega) = \frac{1}{2\pi T} \left| \sum_{t=1}^{T} Z_t e^{-it\omega} \right|^2.$$

The following objective function has been used in the literature:

$$W_T(\theta) = \frac{1}{4\pi} \int_{-\pi}^{\pi} \left\{ \log(f_\theta(\omega)) + \frac{I_T(\omega)}{f_\theta(\omega)} \right\} d\omega. \tag{6.6}$$

By minimizing (6.6) with respect to $\theta \in \Theta$, we have

$$\widehat{\theta}_T = \arg \min_{\theta \in \Theta} W_T(\theta).$$

Under suitable conditions, $\widehat{\theta}_T$ is asymptotically normal. See Dahlhaus (1989) for more details.

This section considers using a discretized version of (6.6) of the form

$$\widehat{W}_T(\theta) = \frac{1}{m} \sum_{j=1}^{m} \left\{ \log(f_\theta(\omega_j)) + \frac{I_T(\omega_j)}{f_\theta(\omega_j)} \right\},$$

where $1 \leq m < \frac{T}{2}$. Let $\Omega = [\Delta_1, \Delta_2]$ with $\frac{1}{2} < \Delta_1 < \Delta_2 < 1$. Clearly, we may estimate θ by

$$\widehat{\theta} = (\widehat{G}, \widehat{H}) = \arg \min_{0 < G < \infty, H \in \Omega} \widehat{W}_T(\theta).$$

Before we state the main results of this section in Theorems 6.1 and 6.2 below, we need to introduce the following conditions.

Assumption 6.1. As $\omega \to 0+$

$$f(\omega) \sim G_0\, \omega^{1-2H_0},$$

where $G_0 \in (0, \infty)$ and $H_0 \in [\Delta_1, \Delta_2]$.

Assumption 6.2. In a neighourhood $(0, \delta)$ of the origin, $f(\omega)$ is differentiable and

$$\frac{d}{d\omega} \log(f(\omega)) = O(\omega^{-1}) \text{ as } \omega \to 0+.$$

Assumption 6.3. We have

$$Z_t = \sum_{s=0}^{\infty} a_s \epsilon_{t-s} \text{ with } \sum_{s=0}^{\infty} a_s^2 < \infty,$$

where

$$E[\epsilon_t | F_{t-1}] = 0,\ E[\epsilon_t^2 | F_{t-1}] = 1,\ a.s.,\ t = 0, \pm 1, \ldots,$$

in which $\{F_t\}$ is a sequence of σ-fields generated by $\{\epsilon_s,\ s \leq t\}$, and there exists a random variable ϵ such that $E[\epsilon^2] < \infty$ and for all $\eta > 0$ and some $C > 0$, $P(|\epsilon_t| > \eta) \leq CP(|\epsilon| > \eta)$.

Assumption 6.4. As $T \to \infty$

$$\frac{1}{m} + \frac{m}{T} \to 0.$$

Robinson (1995b) established the following theorem.

Theorem 6.2. *Let Assumptions 6.1–6.4 hold. Then as $T \to \infty$*

$$\widehat{H} \to_p H_0.$$

The proof is the same as that of Theorem 1 of Robinson (1995b). In order to establish the asymptotic normality, we need to modify Assumptions 6.1–6.4.

Assumption 6.5. For some $\delta \in (0, 2]$

$$f(\omega) \sim G_0 \, \omega^{1-2H_0}(1 + O(\omega^\delta)) \quad \text{as} \quad \omega \to 0+,$$

where $G_0 \in (0, \infty)$ and $H_0 \in [\Delta_1, \Delta_2]$.

Assumption 6.6. In a neighbourhood $(0, \delta)$ of the origin, $\alpha(\omega)$ is differentiable and

$$\frac{d\alpha(\omega)}{d\omega} = O\left(\frac{|\alpha(\omega)|}{\omega}\right) \quad \text{as} \quad \omega \to 0+,$$

where $\alpha(\omega) = \sum_{s=0}^{\infty} a_s e^{is\omega}$.

Assumption 6.7. Assumption 6.3 holds and also

$$E[\epsilon_t^2 | F_{t-1}] = \mu_3 \text{ a.s. and } E[\epsilon_t^4] = \mu_4, \ t = 0, \pm 1, \ldots$$

for finite constants μ_3 and μ_4.

Assumption 6.8. As $T \to \infty$

$$\frac{1}{m} + \frac{m^{1+2\delta}(\log(m))^2}{T^{2\delta}} \to 0.$$

Robinson (1995b) also established the following theorem.

Theorem 6.3. *Under Assumptions 6.5–6.8, we have as $T \to \infty$*

$$\sqrt{m}(\widehat{H} - H_0) \to N\left(0, \frac{1}{4}\right).$$

The proof is the same as that of Theorem 2 of Robinson (1995b).

6.3 Simultaneous semiparametric estimation

As assumed in model (6.5), the above section looks only at the Gaussian semiparametric estimation of the parameters (G, H). This section proposes using a simultaneous semiparametric estimation procedure for a vector of unknown parameters involved in a class of continuous–time Gaussian models. Similarly to Theorems 6.2 and 6.3, both asymptotic consistency and asymptotic normality results are established.

6.3.1 Gaussian models with LRD and intermittency

As an extension to model (6.5), Gao (2004) considered the case where the spectral density function of Gaussian processes is of the form

$$\psi(\omega) = \psi(\omega, \theta) = \frac{\pi(\omega, \theta)\sigma^2}{|\omega|^{2\beta}(\omega^2 + \alpha^2)^\gamma}, \quad \omega \in (-\infty, \infty), \qquad (6.7)$$

where $\theta = (\alpha, \beta, \sigma, \gamma) \in \Theta$ with

$$\Theta = \left\{ 0 < \alpha < \infty, 0 < \beta < \frac{1}{2}, 0 < \sigma < \infty, 0 < \gamma < \infty, \beta + \gamma > \frac{1}{2} \right\},$$

α is normally involved in the drift function of the process involved, β is the LRD parameter, σ is involved in the diffusion function of the process considered, γ is normally called the intermittency parameter of the process considered, and $\pi(\omega, \theta)$ is a continuous and positive function satisfying $0 < \lim_{\omega \to 0}$ or $_{\omega \to \pm\infty} \pi(\omega, \theta) < \infty$ for each $\theta \in \Theta$.

When $\pi(\omega, \theta) \equiv 1$ and $\alpha = 1$ in (6.7), the existence of such a process has been justified in Anh, Angulo and Ruiz–Medina (1999). For this case, model (6.7) corresponds to the fractional Riesz–Bessel motion (fRBm) case. The significance of fRBm is in its behaviour when $|\omega| \to \infty$. It is noted that when $\alpha = 1$, $\psi(\omega)$ of (6.7) is well defined as $|\omega| \to \infty$ due to the presence of the component $(1 + \omega^2)^{-\gamma}$, $\gamma > 0$, which is the Fourier transform of the Bessel potential. As a result, the covariances $R(t)$ of the increments of fRBm are strong for small $|t|$. That is, large (resp. small) values of the increments tend to be followed by large (resp. small) values with probability sufficiently close to one. This is the clustering phenomenon observed in stochastic finance (see Shiryaev 1999, page 365). This phenomenon is referred to as (second-order) intermittency in the turbulence literature (see Frisch 1995).

When $\pi(\omega, \theta) = \frac{1}{\Gamma^2(1+\beta)}$ and $\gamma = 1$, model (6.7) reduces to

$$\psi(\omega) = \psi(\omega, \theta) = \frac{\sigma^2}{\Gamma^2(1 + \beta)} \frac{1}{|\omega|^{2\beta}} \frac{1}{\omega^2 + \alpha^2}, \qquad (6.8)$$

which is just the spectral density of processes that are solutions of continuous–time fractional stochastic differential equations of the form

$$dZ(t) = -\alpha Z(t)dt + \sigma dB_\beta(t), \quad Z(0) = 0, \ t \in (0, \infty), \qquad (6.9)$$

where $B_\beta(t)$ is general fractional Brownian motion given by $B_\beta(t) = \int_0^t \frac{(t-s)^\beta}{\Gamma(1+\beta)} dB(s)$, $B(t)$ is standard Brownian motion, and $\Gamma(x)$ is the usual Γ function. Obviously, model (6.9) is a fractional stochastic differential equation. It is noted that the solution of (6.9) is given by

$$Z(t) = \int_0^t A(t-s)dB(t) \text{ with } t \in [0, \infty) \text{ and}$$

$$A(x) = \frac{\sigma}{\Gamma(1+\beta)} \left(x^\beta - \alpha \int_0^x e^{-\alpha(x-u)} u^\beta du \right). \quad (6.10)$$

Model (6.8) corresponds to the spectral density of an Ornstein–Ulhenbeck process of the form (6.9) driven by fractional Brownian motion with Hurst index $H = \beta + \frac{1}{2}$. Obviously, the process $Z(t)$ of (6.10) is Gaussian.

It is noted that the $\psi(\omega)$ of (6.7) is well defined for both $|\omega| \to 0$ and $|\omega| \to \infty$ due to the presence of the component $(\alpha^2 + \omega^2)^{-\gamma}$, which provides some additional information for the identification and estimation of α and γ. For model (6.8), when $|\omega| \to 0$, $\psi(\omega) \sim \frac{1}{\Gamma^2(1+\beta)} \left(\frac{\sigma}{\alpha} \right)^2 \frac{1}{\omega^{2\beta}}$. For this case, if only information for LRD is used, it is easy to estimate the whole component $\left(\frac{\sigma}{\alpha} \right)^2$ but difficult to estimate both σ and α individually. Thus, the use of information for LRD only can cause a model misspecification problem. This suggests using some additional information for the high frequency area (i.e., $|\omega| \to \infty$) for the identification and estimation of both α and γ involved in model (6.7).

It should be pointed out that the processes having a spectral density of the form (6.7) can be nonstationary. As can be seen from (6.10), $Z(t)$ of (6.10) is a nonstationary Gaussian process, but the spectral density $\psi(\omega)$ is a special case of from (6.7). It is worthwhile to point out that model (6.7) extends and covers many important cases, including the important case where $0 < \beta < \frac{1}{2}$ and $\gamma \geq \frac{1}{2}$. For this case, $\beta + \gamma > \frac{1}{2}$ holds automatically. Recently, Gao *et al.* (2001) considered the special case where $\pi(\omega, \theta) \equiv 1$, $\alpha = 1$, $0 < \beta < 1/2$ and $\gamma \geq 1/2$ in (6.7). The authors were able to establish asymptotic results for estimators of θ based on discretization. See, for example, Theorem 2.2 of Gao *et al.* (2001). As a special case of model (6.7), another important case where $\pi(\omega, \theta) \equiv 1$, $\alpha = 1$, $0 < \beta < \frac{1}{2}$, $0 < \gamma < \frac{1}{2}$ but $\beta + \gamma > \frac{1}{2}$ that was discussed in detail by Gao (2004). There are two reasons to explain why the latter case is quite important. The first reason is that it is theoretically much more difficult to estimate both β and γ when they relate each other in the form of $\beta + \gamma > \frac{1}{2}$. As can be seen from the next section, a constrained estimation procedure is needed for this case. The second reason is that one needs to consider the case where both the long–range dependence and intermittency are moderate but the collective impact of the two is quite significant.

In the following section, we propose using a semiparametric estimation procedure for the parameters involved in (6.7) through using a continuous–time version of the Gauss–Whittle objective function. Both the consistency and the asymptotic normality of the estimators of the parameters are established.

6.3.2 Semiparametric spectral density estimation

Since the process $Z(t)$ of (6.10) is not stationary, we denote by $Y(t)$ the stationary version of $Z(t)$,

$$Y(t) = \int_{-\infty}^{t} A(t - s)dB(s), \ t \in [0, \infty). \tag{6.11}$$

Define the autocovariance function of Y by $\gamma_Y(h) = \text{cov}[Y(t), Y(t + h)]$ for any $h \in (-\infty, \infty)$. In the frequency domain, as the spectral density of $Y(t)$ is idential to that of the process $Z(t)$ (see Proposition 6 of Comte and Renault 1996), the spectral density of $Z(t)$ defined by the Fourier transform of $\gamma_Y(h)$: $\psi(\omega) = \int_{-\infty}^{\infty} e^{-i\omega\tau}\gamma_Y(\tau)d\tau$, is given by

$$\psi(\omega) = \psi(\omega, \theta) = \frac{1}{\Gamma^2(1 + \beta)} \frac{\sigma^2}{|\omega|^{2\beta}(\omega^2 + \alpha^2)^\gamma}, \ \omega \in (-\infty, \infty), \tag{6.12}$$

where $\theta = (\alpha, \beta, \sigma, \gamma)$ is the same as in (6.7). Thus, we may interpret that the spectral density of the form (6.12) corresponds to $Z(t)$ of (6.8). Note that when $\gamma = 1$, the spectral density of (6.12) reduces to form (6.8).

We assume without loss of generality that both $Z(t)$ and $Y(t)$ are defined on $[0, \infty)$. It is easily seen that the proposed estimation procedure remains true when both $Z(t)$ and $Y(t)$ are defined on $(-\infty, \infty)$. This section considers only the case of $0 < \beta < \frac{1}{2}$. For any given $\omega \in (-\infty, \infty)$, we define the following estimator of $\psi(\omega) = \psi(\omega, \theta)$ by

$$I_N^Y(\omega) = \frac{1}{2\pi N} \left| \int_0^N e^{-i\omega t}Y(t)dt \right|^2, \tag{6.13}$$

where $N > 0$ is the upper bound of the interval $[0, N]$, on which each $Y(t)$ is observed. Throughout this chapter, the stochastic integrals are limits in mean square of appropriate Riemann sums. It is noted that form (6.13) for the continuous–time case is an extension of the usual periodogram for the discrete case (see Brockwell and Davis 1990). For discrete time processes, some asymptotic results have already been established for periodogram estimators (see §10 of Brockwell and Davis 1990).

Before establishing the main results of this section, we need to introduce the following assumption.

Assumption 6.9. (i) Assume that each Gaussian process having a spectral density of the form (6.7) has a stationary Gaussian version.

(ii) Assume that $\pi(\omega, \theta)$ is a positive and continuous function in both ω and θ, bounded away from zero and chosen to satisfy

$$\int_{-\infty}^{\infty} \psi(\omega, \theta) \, d\omega < \infty \text{ and } \frac{\partial}{\partial \theta} \left(\int_{-\infty}^{\infty} \log\left(\psi(\omega, \theta)\right) \frac{d\omega}{1 + \omega^2} \right) = 0 \text{ for } \theta \in \Theta.$$

In addition, $\pi(\omega, \theta)$ is a symmetric function in ω satisfying

$$0 < \lim_{\omega \to 0} \pi(\omega, \theta^*) < \infty \text{ and } 0 < \lim_{\omega \to \pm\infty} \pi(\omega, \theta^*) < \infty$$

for each given $\theta^* \in \Theta$.

(iii) Let θ_0 be the true value of θ, and θ_0 be in the interior of Θ_0, a compact subset of Θ. For any small $\epsilon > 0$, if $\epsilon < ||\theta - \theta_0|| < \frac{1}{4}$ then

$$\int_{-\infty}^{\infty} \frac{\psi(\omega, \theta_0)}{\psi(\omega, \theta)} \frac{1}{1 + \omega^2} \, d\omega < \infty,$$

where $|| \cdot ||$ denotes the Euclidean norm.

Assumption 6.9(i) assumes only that the processes having a spectral density of the form (6.7) are Gaussian processes, which can be solutions of fractional stochastic differential equations. For example, the process $Z(t)$ given in (6.10) is the solution of equation (6.9), and the spectral density of the solution is given by (6.8). Assumption 6.9(ii) assumes that $\psi(\omega, \theta)$ is also normalized so that

$$\frac{\partial}{\partial \theta} \left(\int_{-\infty}^{\infty} \log\left(\psi(\omega, \theta)\right) \frac{d\omega}{1 + \omega^2} \right) = 0.$$

This extends similar conditions introduced by Fox and Taqqu (1986) and then generalized by Heyde and Gay (1993).

Assumption 6.9 allows a lot of flexibility in choosing the form of $\pi(\omega, \theta)$, which includes not only the LRD parameter β, but also the parameters–of–interest, α and σ. The last two parameters, as can be seen from models (6.8) and (6.9), have some financial interpretations: α represents the speed of the fluctuations of an interest rate data set while σ is a measure for the order of the magnitude of the fluctuations of an interest rate data set around zero, for example. In general, $\pi(\omega, \theta)$ represents some kind of magnitude of the process involved. Assumption 6.1 holds in many cases. For example, when $\pi(\omega, \theta) = \frac{1}{\Gamma^2(1+\beta)}$ and $\gamma \equiv 1$ or $\pi(\omega, \theta) \equiv 1$, Assumption 6.9 holds automatically.

We propose using a simultaneous Gaussian semiparametric estimation procedure based on the following objective function:

$$L_N^Y(\theta) = \frac{1}{4\pi} \int_{-\infty}^{\infty} \left\{ \log(\psi(\omega, \theta)) + \frac{I_N^Y(\omega)}{\psi(\omega, \theta)} \right\} \frac{d\omega}{1 + \omega^2}. \qquad (6.14)$$

The weight function $\frac{1}{1+\omega^2}$ involved is to ensure that $L_N^Y(\theta)$ is well defined. This is mainly because $\lim_{\omega \to \pm \infty} \frac{\log(\psi(\omega,\theta))}{1+\omega^2} = 0$.

Due to the form of

$$\Theta = \left\{ \theta : 0 < \alpha < \infty, 0 < \beta < \frac{1}{2}, 0 < \sigma < \infty, 0 < \gamma < \infty, \beta + \gamma > \frac{1}{2} \right\},$$

we need to consider the following two different cases:

- Case I:

$$\Theta_1 = \left\{ \theta : 0 < \alpha < \infty, \ 0 < \sigma < \infty, \ 0 < \beta < \frac{1}{2}, \frac{1}{2} \le \gamma < \infty \right\};$$

- Case II:

$$\Theta_2 = \left\{ \theta : 0 < \alpha < \infty, \ 0 < \sigma < \infty, \ 0 < \beta, \gamma < \frac{1}{2}, \ \beta + \gamma > \frac{1}{2} \right\}.$$

Obviously, $\Theta_1 \subset \Theta$ and $\Theta_2 \subset \Theta$.

For Case I, the minimum contrast estimator of θ is defined by

$$\bar{\theta}_N = \arg \min_{\theta \in \Theta_{10}} L_N^Y(\theta), \qquad (6.15)$$

where Θ_{10} is a compact subset of Θ_1.

For Case II, we introduce the following Lagrangian function

$$M_N^Y(\theta) = L_N^Y(\theta) - \lambda g(\theta),$$

where λ is the multiplier and $g(\theta) = \beta + \gamma - \frac{1}{2}$. The minimisation problem:

$$\text{Minimising } L_N^Y(\theta), \text{ subject to } \theta \in \Theta_2$$

can now be transferred to the following minimisation problem:

$$\tilde{\theta}_N = \arg \min_{\theta \in \Theta_{20}} M_N^Y(\theta), \qquad (6.16)$$

where Θ_{20} is a compact subset of Θ_2. It should be noted that Case I corresponds to $\lambda = 0$ and that Case II corresponds to $\lambda \neq 0$. To avoid abusing the notation of θ_0, we denote the true value of $\theta \in \Theta_1$ by θ_{10}, and the true value of $\theta \in \Theta_2$ by θ_{20} throughout the rest of this section.

To state the following results, we also need to introduce the following conditions.

Assumption 6.10. (i) For any real function $p(\cdot, \cdot) \in L^2(-\infty, \infty)$,

$$\int_{-\infty}^{\infty} \frac{p^2(\omega, \theta_0)}{(1 + \omega^2)^2} \left(\frac{\partial \log(\psi(\omega, \theta))}{\partial \theta} \right)^\tau \left(\frac{\partial \log(\psi(\omega, \theta))}{\partial \theta} \right) |_{\theta = \theta_0} \, d\omega < \infty,$$

where $\theta_0 = \theta_{10}$ or θ_{20}.

(ii) For $\theta \in \Theta$,

$$\Sigma(\theta) = \frac{1}{4\pi} \int_{-\infty}^{\infty} \left(\frac{\partial \log(\psi(\omega, \theta))}{\partial \theta} \right) \left(\frac{\partial \log(\psi(\omega, \theta))}{\partial \theta} \right)^\tau \frac{1}{(1 + \omega^2)^2} \, d\omega < \infty.$$

(iii) The inverse matrix, $\Sigma^{-1}(\theta_0)$, of $\Sigma(\theta_0)$ exists, where $\theta_0 = \theta_{10}$ or θ_{20}.

Assumption 6.11. Assume that $K(\theta, \theta_0)$ is convex in θ on an open set $\mathcal{C}(\theta_0)$ containing θ_0, where $\theta_0 = \theta_{10}$ or θ_{20} and

$$K(\theta, \theta_0) = \frac{1}{4\pi} \int_{-\infty}^{\infty} \left\{ \frac{\psi(\omega, \theta_0)}{\psi(\omega, \theta)} - 1 - \log \left(\frac{\psi(\omega, \theta_0)}{\psi(\omega, \theta)} \right) \right\} \frac{d\omega}{1 + \omega^2}.$$

Assumption 6.10(i) is required for an application of a continuous–time central limit theorem to the proof of the asymptotic normality. Assumption 6.10(ii)(iii) is similar to those for the discrete case. See, for example, Condition (A2) of Heyde and Gay (1993). Assumptions 6.10 and 6.11 simplify some existing conditions for continuous–time models. See, for example, Conditions 2.1 and 2.2 of Gao, Anh and Heyde (2002).

Assumption 6.10 holds in many cases. For example, when $\pi(\omega, \theta) = \frac{1}{\Gamma^2(1+\beta)}$ and $\gamma \equiv 1$ or $\pi(\omega, \theta) \equiv 1$, Assumption 6.10 holds automatically. It should be pointed out that Assumption 6.11 holds automatically for the case where $\pi(\omega, \theta) \equiv 1$, as the matrix $\mathcal{K}(\theta) = \{k_{ij}(\theta)\}_{\{1 \leq i,j \leq 4\}}$ is positive semidefinite for every $\theta \in \mathcal{C}(\theta_0)$, an open convex set containing θ_0, where $k_{ij}(\theta) = \frac{\partial^2}{\partial \theta_i \partial \theta_j} K(\theta, \theta_0)$, in which $\theta_1 = \alpha$, $\theta_2 = \beta$, $\theta_3 = \sigma$ and $\theta_4 = \gamma$. For the detailed verification, we need to use Theorem 4.5 of Rockafeller (1970). This suggests that Assumption 6.11 is a natural condition.

In general, in order to ensure the existence and uniqueness (at least asymptotically) of $\widehat{\theta}_N$, the convexity imposed in Assumption 6.11 is necessary. Previously, this type of condition has not been mentioned in detail, mainly because the convexity condition holds automatically in some special cases. For our model (6.7), as the form of $\psi(\omega, \theta)$ is very general, Assumption 6.11 is needed for rigorousness consideration.

We now state the following results for Case I and Case II in Theorems 6.4 and 6.5, respectively.

Theorem 6.4 (Case I). (i) *Assume that Assumptions 6.9–6.11 with* $\theta_0 = \theta_{10}$ *hold. Then*

$$P\left(\lim_{N \to \infty} \bar{\theta}_N = \theta_{10}\right) = 1.$$

(ii) *In addition, if the true value* θ_{10} *of* θ *is in the interior of* Θ_{10}, *then as* $N \to \infty$

$$\sqrt{N}(\bar{\theta}_N - \theta_{10}) \to_D N\left(0, \Sigma^{-1}(\theta_{10})\right),$$

where $\Sigma^{-1}(\theta_{10})$ *is as defined above.*

Theorem 6.5 (Case II). (i) *Assume that Assumptions 6.9–6.10 with* $\theta_0 = \theta_{20}$ *hold. In addition, let* $\tilde{\theta}_N$ *converge to* θ_{20} *with probability one and the true value* θ_{20} *of* θ *be in the interior of* Θ_{20}. *Then as* $N \to \infty$

$$\sqrt{N}(\tilde{\theta}_N - \theta_{20}) \to_D N\left(0, A\Sigma^{-1}(\theta_{20})A\right),$$

where the 4×4 *matrix* A *is given by*

$$A = \begin{pmatrix} 1 & 0 & 0 & 0 \\ 0 & \frac{1}{2} & 0 & -\frac{1}{2} \\ 0 & 0 & 1 & 0 \\ 0 & -\frac{1}{2} & 0 & \frac{1}{2} \end{pmatrix}.$$

(ii) *Assume that Assumptions 6.9–6.11 with* $\theta_0 = \theta_{20}$ *hold. Then*

$$P\left(\lim_{N \to \infty} \tilde{\theta}_N = \theta_{20}\right) = 1.$$

The proofs of Theorems 6.4 and 6.5 are available from Gao (2004).

Theorem 6.4 extends and complements some existing results for both the discrete and continuous time cases. See, for example, Comte and Renault (1996, 1998), Gao *et al.* (2001), and Gao, Anh and Heyde (2002). As can be seen, strong consistency and asymptotic normality results of the estimators of the parameters involved in (6.7) do not depend on the use of discretised values of the process under consideration. It should also be pointed out that the use of continuous–time models can avoid the problem of misspecification for parameters. Moreover, the estimation procedure fully makes the best use of all the information available and therefore can clearly identify and estimate all the four parameters involved.

Theorem 6.5 establishes the asymptotic consistency results for the case where $\theta \in \Theta_2$. The corresponding estimation procedure for the important class of models is now applicable to the case where the LRD parameter β satisfies $0 < \beta < \frac{1}{2}$, the intermittency parameter γ satisfies $0 < \gamma < \frac{1}{2}$, but the pair (β, γ) satisfies the condition: $\beta + \gamma > \frac{1}{2}$. Some practical problems that have not solved previously can now be dealt with. We need to point out that the strong consistency of $\hat{\theta}_N$ is necessary for the establishment of the asymptotic normality and that Assumption 6.11 may only be one of the few necessary conditions for the strong consistency. Due to this reason, we impose the strong consistency directly for the establishment of the asymptotic normality.

In the following section, we have a detailed look at an application of the proposed estimation procedure to a class of continuous–time long–range dependent stochastic volatility models exactly as discussed in Casas and Gao (2005).

6.4 LRD stochastic volatility models

6.4.1 Introduction

More than thirty years ago, Black and Scholes (1973) assumed a constant volatility to derive their famous option pricing equation. The implied volatility values obtained from this equation show skewness, suggesting that the assumption of constant volatility is not feasible. In fact, the volatility shows an intermittent behaviour with periods of high values and periods of low values. In addition, the asset volatility cannot be directly observed. Stochastic volatility (SV) models deal with these two facts. Hull and White (1987) were amongst the first to study the logarithm of the stochastic volatility as an Ornstein–Uhlenbeck process. A review and comparative study about modeling SV up to 1994 has been given by Taylor (1994). Andersen and Sørensen (1996) examined generalized moments of method for estimating stochastic volatility model. Andersen and Lund (1997) extended the CIR model to associate the spot interest rate with stochastic volatility process through estimating the parameters with the efficient method of moments. The main assumption of the SV model is that the volatility is a lognormal process. The probabilistic and statistical properties of a lognormal are well known. However, the parametric estimation has not been uncomplicated due to the difficulty finding the maximum likelihood (ML) function. Since 1994, estimation procedures have been proposed. A comprehensive survey on several different estimation procedures developed for the SV model has been given in Broto and Ruiz (2004).

Recent studies show that some data may display long–range dependence (LRD) (see the detailed review by Beran 1994; Baillie and King 1996; Anh and Heyde 1999; and Robinson 2003). Since about ten years ago, there has been some work on studying stochastic volatility with LRD.

Breidt, Crato and de Lima (1998), Comte and Renault (1998), and Harvey (1998) were among the first to consider long–memory stochastic volatility (LMSV) models. Breidt, Crato and de Lima (1998) also considered an LMSV case where the log–volatility is modelled as an ARFIMA process. Comte and Renault (1998) consider a continuous–time fractionally stochastic volatility (FSV) model of the form

$$dY(t) = v(t)dB_1(t) \quad \text{and} \quad dx(t) = -\alpha x(t)dt + \sigma dB_\beta(t), \qquad (6.17)$$

where $x(t) = \ln(v(t))$, $Y(t) = \ln(S(t))$ with $S(t)$ being the return process, $B_1(t)$ is a standard Brownian motion, α is the drift parameter, $\sigma > 0$ is the volatility parameter, and $B_\beta(t)$ is a fractionally Brownian motion process of the form: $B_\beta(t) = \int_0^t \frac{(t-s)^\beta}{\Gamma(1+\beta)} dB(s)$, in which $B(t)$ is a standard Brownian motion and $\Gamma(x)$ is the usual Γ function. It is assumed that $B_1(t)$ and $B_\beta(t)$ are independent for all $-\frac{1}{2} < \beta < \frac{1}{2}$.

In Comte and Renault (1998), a discretization procedure was first proposed to approximate the solution of their continuous–time FSV model. An estimation procedure for $0 < \beta < \frac{1}{2}$ is developed for a discretized version of the solution $Y(t)$ based on the so–called log–periodogram regression. Deo and Hurvich (2001) further studied such an estimation procedure based on the log–periodogram regression method. The authors systematically established the mean–squared error properties as well as asymptotic consistency results for an estimator of β. Section 4 of Broto and Ruiz (2004) provides a good survey about existing estimation methods in discrete–time LMSV models. Gao (2004) pointed out that it is possible to estimate all the parameters involved in model (6.17) using the so–called continuous–time version of the Gauss–Whittle contrast function method proposed in Gao et al. (2001) and Gao, Anh and Heyde (2002).

The paper by Casas and Gao (2006) has considered a general class of stochastic volatility models of the form

$$dY(t) = V(t)dB_1(t) \qquad (6.18)$$

with $V(t)$ being given by $V(t) = e^{X(t)}$, where

$$X(t) = \int_{-\infty}^t A(t - s)dB(s), \qquad (6.19)$$

in which $B_1(t)$ and $B(t)$ are two standard Brownian motion processes, $A(\cdot)$ is a deterministic function such that $X(t)$ is stationary, and the explicit expression of $A(\cdot)$ is determined by the spectral density of $X(t)$. In addition, the authors assumed that the spectral density function is given by

$$\psi_X(\omega) = \psi_X(\omega, \theta) = \frac{\pi(\omega, \theta)\sigma^2}{|\omega|^{2\beta}\,(\omega^2 + \alpha^2)}, \quad \omega \in (-\infty, \infty), \tag{6.20}$$

where

$$\theta = (\alpha, \beta, \sigma) \in \Theta = \left\{0 < \alpha < \infty, -\frac{1}{2} < \beta < \frac{1}{2}, 0 < \sigma < \infty\right\},$$

$\pi(\omega, \theta)$ is a either a parametric function of θ or a semiparametric function of θ and an unknown function, continuous and positive function satisfying $0 < \lim_{\omega \to 0}$ or $_{\omega \to \pm\infty}\, \pi(\omega, \theta) < \infty$ for each given $\theta \in \Theta$, α is normally involved in the drift function of the stochastic volatility process, β is the LRD parameter and σ is a kind of volatility of the stochastic volatility process.

Unlike most existing studies assuming a particular form for the volatility process $V(t)$, Casas and Gao (2006) have implicitly imposed certain conditions on the distributional structure of the volatility process. First, the volatility is a lognormal process with a vector of parameters involved. Second, the vector of parameters is specified through a corresponding spectral density function. Third, the parameters involved in the spectral density function may be explicitly interpreted and fully estimated. Fourth, the generality of the spectral density function of the form (6.20) implicitly implies that the class of lognormal volatility processes can be quite general. As a matter of the fact, the class of models (6.18)–(6.20) is quite general to cover some existing models. For example, when $\pi(\omega, \theta) = \frac{1}{\Gamma^2(1+\beta)}$, model (6.20) reduces to

$$\psi_X(\omega) = \psi_X(\omega, \theta) = \frac{\sigma^2}{\Gamma^2(1+\beta)} \frac{1}{|\omega|^{2\beta}} \frac{1}{\omega^2 + \alpha^2}, \tag{6.21}$$

which is just the spectral density of the solutions of the second equation of (6.17) given by

$$x(t) = \int_0^t A(t-s)dB(t) \quad \text{with}$$

$$A(x) = \frac{\sigma}{\Gamma(1+\beta)} \left(x^\beta - \alpha \int_0^x e^{-\alpha(x-u)} u^\beta du\right). \tag{6.22}$$

Its stationary version is defined as $X(t) = \int_{-\infty}^t A(t-s)dB(t)$. Existence of some other models corresponding to (6.19) has been established by Anh and Inoue (2005) and Anh, Inoue and Kasahara (2005). Instead of further discussing such existence, Casas and Gao (2006) have concentrated on the parametric estimation of the volatility process (6.19) by estimating the parameters involved in the spectral density function

(6.20). The authors then demonstrated how to implement the proposed estimation procedure in practice by using both simulated and real sets of data.

The main structure of Section 6.4 can be summarized as follows: (i) it considers a general class of stochastic volatility models with either LRD, intermediate range dependence (IRD), or short–range dependence (SRD); (ii) it proposes an estimation procedure to deal with cases where a class of non–Gaussian processes may display LRD, IRD or SRD; (iii) some comprehensive simulation studies show that the proposed estimation procedure works well numerically not only for the LRD parameter β, but also for both the drift parameter α and the variance σ^2; and (iv) the methodology is also applied to the estimation of the volatility of several well–known stock market indexes.

6.4.2 Simultaneous semiparametric estimation

Casas and Gao (2006) have proposed an estimation procedure based on discrete observations of $Y(t)$ in (6.18). This is mainly because observations on $Y(t)$ are made at discrete intervals of time in many practical circumstances, even though the underlying process may be continuous.

Consider a discretized version of model (6.18) of the form

$$Y_{t\Delta} - Y_{(t-1)\Delta} = V_{(t-1)\Delta}(B_{t\Delta} - B_{(t-1)\Delta}), \ t = 1, 2, \cdots, T, \qquad (6.23)$$

where Δ is the time between successive observations and T is the size of observations. In theory, we may study asymptotic properties for our estimation procedure for either the case where Δ is small but fixed or the case where Δ is varied according to T. We focus on the case where Δ is small but fixed throughout the rest of this section, since this section is mainly interested in estimating stochastic volatility process $V(t)$ using either monthly, weekly, daily, or higher frequency returns.

Let $W_t = \frac{Y_{t\Delta} - Y_{(t-1)\Delta}}{\Delta}$, $U_t = V_{(t-1)\Delta}$ and $\epsilon_t = \sqrt{\Delta^{-1}}(B_{t\Delta} - B_{(t-1)\Delta})$. Then model (6.23) may be rewritten as

$$W_t = U_t \sqrt{\Delta^{-1}} \ \epsilon_t, \ t = 1, 2, \cdots, T, \qquad (6.24)$$

where $\{\epsilon_t\}$ is a sequence of independent and identically distributed (i.i.d.) normal errors drawn from $N(0, 1)$, and $\{\epsilon_t\}$ and $\{U_s\}$ are mutually independent for all $s, t \geq 1$. Letting $Z_t = \log(W_t^2)$, $X_t = \log(U_t)$, $e_t = \log(\epsilon_t^2) - E[\log(\epsilon_t^2)]$ and $\mu = E[\log(\epsilon_t^2)] - \log(\Delta)$, model (6.24) implies that

$$Z_t = \mu + 2X_t + e_t, \ t = 1, 2, \cdots, T, \qquad (6.25)$$

where $\{X_t\}$ is a stationary Gaussian time series with LRD, and $\{e_t\}$ is a sequence of i.i.d. random errors with a known distributional structure.

Such a simple linear model has been shown to be pivotal for establishing various consistent estimation procedures (Deo and Hurvich 2001). Our estimation procedure is also based on model (6.25). Since both Z_t and X_t are stationary, their corresponding spectral density functions $f_Z(\cdot, \cdot)$ and $f_X(\cdot, \cdot)$ satisfy the following relationship:

$$f_Z(\omega, \theta) = 4 f_X(\omega, \theta) + \frac{\sigma_e^2}{2\pi} = 4 f_X(\omega, \theta) + \frac{\pi}{4}, \qquad (6.26)$$

where $\sigma_e^2 = \frac{\pi^2}{2}$ is used in (6.26).

Since $\{X_t\}$ is a sequence of discrete observations of the continuous–time process $X(t)$, existing results (Bloomfield 1976, §2.5) imply that the spectral density function $f_X(\omega, \theta)$ is expressed as

$$f_X(\omega, \theta) = \frac{1}{\Delta} \sum_{k=-\infty}^{\infty} \psi_X \left(\frac{\omega - 2k\pi}{\Delta}, \theta \right), \qquad (6.27)$$

which, together with (6.26), implies that the spectral density function of Z_t is given by

$$f_Z(\omega, \theta) = \frac{4}{\Delta} \sum_{k=-\infty}^{\infty} \psi_X \left(\frac{\omega - 2k\pi}{\Delta}, \theta \right) + \frac{\pi}{4}. \qquad (6.28)$$

The spectral density $f(\omega, \theta)$ is estimated by the following periodogram

$$I_T(\omega) = I_T^Z(\omega) = \frac{1}{2\pi T} \left| \sum_{t=1}^{T} e^{-i\omega t} Z_t \right|^2. \qquad (6.29)$$

The following Whittle contrast function is then employed

$$W_T(\theta) = \frac{1}{4\pi} \int_{-\pi}^{\pi} \left\{ \log(f_Z(\omega, \theta)) + \frac{I_T(\omega)}{f_Z(\omega, \theta)} \right\} d\omega \qquad (6.30)$$

and θ is estimated by

$$\widehat{\theta}_T = \arg \min_{\theta \in \Theta_0} W_T(\theta), \qquad (6.31)$$

where Θ_0 is a compact subset of the parameter space Θ.

Equations (6.29)–(6.31) have been working well both in theory and practice for the case where the underlying process $\{Z_t\}$ is Gaussian. Our theory and simulation results below show that such an estimation procedure also works well both theoretically and practically for the case where $\{Z_t\}$ is stationary but non–Gaussian.

To state the main theoretical results of this section, the following assumptions are needed. For simplicity, denote $\theta = (\alpha, \beta, \sigma)^\tau = (\theta_1, \theta_2, \theta_3)^\tau$.

Assumption 6.12. (i) Consider the general model structure given by (6.18)–(6.20). Suppose that the two standard Brownian motion processes $B_1(t)$ and $B(s)$ are mutually independent for all $-\infty < s, t < \infty$.

(ii) Assume that $\pi(\omega, \theta)$ is a positive and continuous function in both ω and θ, bounded away from zero and chosen to satisfy

$$\int_{-\pi}^{\pi} f_X(\omega, \theta) \, d\omega < \infty \quad \text{and} \quad \int_{-\pi}^{\pi} \log\left(f_X(\omega, \theta)\right) d\omega > -\infty \quad \text{for} \quad \theta \in \Theta.$$

In addition, $\pi(\omega, \theta)$ is a symmetric function in ω satisfying

$$0 < \lim_{\omega \to 0} \pi(\omega, \theta^*) < \infty \quad \text{and} \quad 0 < \lim_{\omega \to \pm\infty} \pi(\omega, \theta^*) < \infty$$

for each given $\theta^* \in \Theta$.

(iii) Assume that $L(\theta, \theta_0)$ is convex in θ on an open set $\mathcal{C}(\theta_0)$ containing θ_0, where

$$L(\theta, \theta_0) = \frac{1}{4\pi} \int_{-\pi}^{\pi} \left\{ \frac{f_Z(\omega, \theta_0)}{f_Z(\omega, \theta)} - 1 - \log\left(\frac{f_Z(\omega, \theta_0)}{f_Z(\omega, \theta)}\right) \right\} d\omega.$$

Assumption 6.13. The functions $f_X(\omega, \theta)$ and $g_{iX}(\omega, \theta) = -\frac{\partial f_X^{-1}(\omega, \theta)}{\partial \theta_i}$ for $1 \leq i \leq 3$ satisfy the following properties:

(i) $\int_{-\pi}^{\pi} \log(f_X(\omega, \theta)) \, d\omega$ is twice differentiable in θ under the integral sign;

(ii) $f_X(\omega, \theta)$ is continuous at all $\omega \neq 0$ and $\theta \in \Theta$, $f_X^{-1}(\omega, \theta)$ is continuous at all (ω, θ);

(iii) the inverse function $f_X^{-1}(\omega, \theta)$, $\omega \in (-\pi, \pi]$, $\theta \in \Theta$, is twice differentiable with respect to θ and the functions $\frac{\partial}{\partial \theta_i} f_X^{-1}(\omega, \theta)$ and $\frac{\partial^2}{\partial \theta_j \partial \theta_k} f_X^{-1}(\omega, \theta)$ are continuous at all (ω, θ), $\omega \neq 0$, for $1 \leq i, j, k \leq 3$;

(iv) the functions $g_{iX}(\omega, \theta)$ for $1 \leq i \leq 3$ are symmetric about $\omega = 0$ for $\omega \in (-\pi, \pi]$ and $\theta \in \Theta$;

(v) $g_{iX}(\omega, \theta) \in L_1((-\pi, \pi])$ for all $\theta \in \Theta$ and $i = 1, 2, 3$;

(vi) $f_X(\omega, \theta) g_{iX}(\omega, \theta)$ for $1 \leq i \leq 3$ are in $L_1((-\pi, \pi])$ and $L_2((-\pi, \pi])$ for all $\theta \in \Theta$;

(vii) there exists a constant $0 < k \leq 1$ such that $|\omega|^k f_X(\omega, \theta)$ is bounded and $\frac{g_{iX}(\omega, \theta)}{|\omega|^k}$ for $1 \leq i \leq 3$ are in $L_2((-\pi, \pi])$ for all $\theta \in \Theta$; and

(viii) the matrix $\{\frac{\partial}{\partial\theta}\log(f_X(\omega,\theta))\}\{\frac{\partial}{\partial\theta}\log(f_X(\omega,\theta))\}^\tau$ is in $L_1((-\pi,\pi])\times$ Θ_1, where $\Theta_1 \in \Theta$.

Assumption 6.12 imposes some conditions to ensure the identifiability and existence of the spectral density function $f_X(\omega,\theta)$ and thus the Gaussian time series $\{X_t\}$.

Assumption 6.13 requires that the spectral density function $f_X(\omega,\theta)$ needs to satisfy certain smoothness and differentiability conditions in order to verify conditions (A2) and (A3) of Heyde and Gay (1993).

Both Assumptions 6.12 and 6.13 are necessary for us to establish the following asymptotic consistency results. The assumptions are justifiable when the form of $\pi(\omega,\theta)$ is specified. For example, when $\pi(\omega,\theta) = \frac{1}{\Gamma^2(1+\beta)}$ and $\Delta = 1$, Assumptions 6.4 and 6.13 hold automatically. In this case, it is obvious

$$
\int_{-\pi}^{\pi} f_X(\omega,\theta)\,d\omega = 4\int_{-\pi}^{\pi}\left(\sum_{k=-\infty}^{\infty}\psi_X(\omega-2k\pi,\theta)\right)d\omega
$$

$$
= 4\int_{-\infty}^{\infty}\psi_X(\omega,\theta)\,d\omega < \infty.
$$

For the second part of Assumption 6.12(i), using the following decomposition

$$
\begin{aligned}
f_X(\omega,\theta) &= 4\frac{\pi(\omega,\theta)\sigma^2}{|\omega|^{2\beta}\,(\omega^2+\alpha^2)} + 4\sum_{k=1}^{\infty}\frac{\pi(2k\pi-\omega,\theta)\sigma^2}{|2k\pi-\omega|^{2\beta}\,((2k\pi-\omega)^2+\alpha^2)} \\
&\quad + 4\frac{\pi(2k\pi+\omega,\theta)\sigma^2}{|2k\pi+\omega|^{2\beta}\,((2k\pi+\omega)^2+\alpha^2)}, \\
&\equiv f_1(\omega,\theta) + f_2(\omega,\theta) + f_3(\omega,\theta),
\end{aligned}
$$

we have

$$
\int_{-\pi}^{\pi}\log(f_X(\omega,\theta))\,d\omega \geq \int_{-\pi}^{\pi}\log(f_1(\omega,\theta))\,d\omega > -\infty
$$

when $\pi(\omega,\theta)$ is specified as $\pi(\omega,\theta) = \frac{1}{\Gamma^2(1+\beta)}$ and $\Delta = 1$.

For the case of $\pi(\omega,\theta) = \frac{1}{\Gamma^2(1+\beta)}$, Assumption 6.13 may be justified similarly as in the proof of Lemma B.1 of Gao et al. (2001). Instead of giving such detailed verification, we establish the following theorem.

Theorem 6.6. *Suppose that Assumptions 6.12 and 6.13 hold. Then*

(i) $\widehat{\theta}_T$ *is a strongly consistent estimator of* θ_0.

(ii) *Furthermore, if the true value θ_0 is in the interior of Θ_0, then as $T \to \infty$*

$$\sqrt{T}(\widehat{\theta}_T - \theta_0) \to N(0, \Sigma^{-1}(\theta_0)),$$

where

$$\Sigma(\theta) = \frac{1}{4\pi} \int_{-\pi}^{\pi} \left(\frac{\partial}{\partial \theta} \log(f_Z(\omega, \theta)) \right) \left(\frac{\partial}{\partial \theta} \log(f_Z(\omega, \theta)) \right)^{\tau} d\omega.$$

Similarly to some existing results (Heyde and Gay 1993; Robinson 1995a; Deo and Hurvich 2001; and others), Theorem 6.6 shows that $\widehat{\theta}_T$ is still a $\sqrt{T}$–consistent estimator of θ_0 even when $\{Z_t\}$ is non–Gaussian. The proof is based on an application of Theorem 1(ii) of Heyde and Gay (1993) and relegated to Section 6.5 of this chapter.

It should be mentioned that as $T \to \infty$, $\sqrt{\Sigma(\theta_0)\, T} \left(\widehat{\theta}_T - \theta_0 \right)$ converges in distribution to $N(0, I)$ regardless of whether Δ is fixed or varied according to T. This implies that in theory the applicability of the proposed estimation procedure does not depend on the choice of Δ. In practice, $\{Z_t\}$ is sampled using monthly, weekly, daily, or higher frequency data. In the following section, we apply our theory and estimation procedure to model (6.7). Our simulation results show that the proposed theory and estimation procedure works quite well numerically.

6.4.3 Simulation results

Consider a simple model of the form

$$dY(t) = e^{X(t)} dB_1(t) \quad \text{with} \quad X(t) = \int_{-\infty}^{t} A(t - s) dB(s), \qquad (6.32)$$

where $A(x) = \frac{\sigma}{\Gamma(1+\beta)} \left(x^{\beta} - \alpha \int_0^x e^{-\alpha(x-u)} u^{\beta} du \right)$. Note that $X(t)$ is the stationary version of $x(t) = \int_0^t A(t - s) dB(s)$, which is the solution of

$$dx(t) = -\alpha x(t) dt + \sigma dB_{\beta}(t). \qquad (6.33)$$

In order to implement the proposed estimation procedure, we need to generate $\{X_t\}$ from such a Gaussian process with LRD. A closely related simulation procedure is given in Comte (1996), who proposed a discrete approximation to the solution of the continuous–time process $X(t)$. A simulation procedure based on the simulation of the covariance function of $X(t)$ has been proposed in Casas and Gao (2006) and summarized as follows:

- generate C_T, a $T \times T$ auto–covariance matrix, using the auto–covariance

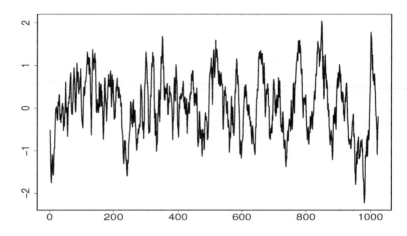

Figure 6.1 *Sample path for data generated with* $\theta = (\alpha, \beta, \sigma) = (0.1, 0.1, 0.1)$ *and* $\Delta = 1$.

function given by $\gamma_X(\tau) = 2 \int_0^\infty f_X(\omega, \theta) \cos(\omega \tau) \, d\omega$ with $\Delta = 1$. C_T is then a symmetric nonnegative definite matrix with spectral decomposition $C_T = V \Lambda V^\top$, where $\Lambda = \text{diag}\{\lambda_1, \ldots, \lambda_T\}$ is the diagonal matrix of the eigenvalues and V is the orthogonal matrix of the eigenvectors such that $V^\tau V = I$ with V^τ being the matrix transpose of V;

- generate a sample $G = (g_1, g_2, \ldots, g_T)^\top$ of independent realisations of a multivariate Gaussian random vector with the zero vector as the mean and the identity matrix as the covariance matrix; and

- generate $(X_1, \ldots, X_T) = V \Lambda^{1/2} V^\tau G$ as the realisation of a multivariate Gaussian random vector with the zero vector as the mean and C_T as the covariance matrix.

The sample path for $\{X_t\}$ generated with the initial parameter values $\theta_0 = (\alpha, \beta, \sigma) = (0.1, 0.1, 0.1)$ is illustrated in Figure 6.1. The periodogram and the spectral density of the simulated data set are illustrated in Figure 6.2.

$\{Z_t\}$ is then generated from (6.25) with $\Delta = 1$. To consider all possible cases, the proposed estimation procedure has been applied to the LRD case of $0 < \beta < \frac{1}{2}$, the IRD case of $-\frac{1}{2} < \beta < 0$ and the case of $\beta = 0$. Sample sizes of $T = 512$ and 1024 were considered. The number of 100 replications was used for each case. Simulation results are displayed in Tables 6.1–6.4 below. The results in Tables 6.1–6.4 show the empirical means, the empirical standard deviations and the empirical mean

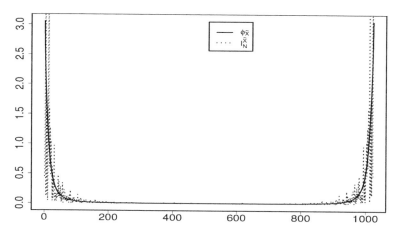

Figure 6.2 *The periodogram and the spectral density for* $\theta = (\alpha, \beta, \sigma) = (0.1, 0.1, 0.1)$ *and* $\Delta = 1$.

squared errors (MSEs) of the estimators. The empirical mean of the absolute value of the corresponding estimated bias is given in each bracket beneath the corresponding estimator. Each of the MSEs is computed as the sum of two terms: the square of the estimated bias and variance. The following tables are taken from Casas and Gao (2006).

Table 6.1 provides the corresponding results for the case where $\theta_0 = (0.8, 0, 0.5)$. These results show that the estimation procedure works well for the SRD case where the initial parameter value for β is zero. For the estimates of both α_0 and σ_0, the empirical mean squared errors (MSEs) decrease when T increases from 512 to 1024. Table 6.2 also considers the case of $\beta_0 = 0$ but with smaller σ_0 and much smaller α_0. For the case of $\theta_0 = (0.001, 0, 0.01)$, the results in the second section show that there is some distortion in the MSEs for the sample sizes of $T = 512$ and 1024. The MSEs become stable and relative smaller than these for the case of $T = 1024$. This supports empirical financial evidence that sufficiently large sample sizes are needed to make precise estimation for very small drift parameters.

From the third section of Table 6.1 to Tables 6.2–6.4, several different pairs of positive and negative β values are considered. The corresponding results show that the MSEs for both α_0 and σ_0 remain stable when β_0 changes from a positive value to its negative counterpart.

Individually, both relatively large and relatively small values of β have been used to assess whether the estimation procedure is sensitive to the

	T = 512			T = 1024		
$\theta_0 = (0.8, 0, 0.5)$	$\hat{\alpha}$	$\hat{\beta}$	$\hat{\sigma}$	$\hat{\alpha}$	$\hat{\beta}$	$\hat{\sigma}$
Empirical mean	0.7776	0.0132	0.5226	0.7923	0.0095	0.5163
	(0.0885)	(0.0623)	(0.0805)	(0.0707)	(0.0453)	(0.0514)
Empirical std.dev.	0.1512	0.0925	0.1130	0.1198	0.0587	0.0690
	(0.1243)	(0.0694)	(0.0821)	(0.0967)	(0.0383)	(0.0487)
Empirical MSE	0.0234	0.0087	0.0133	0.0144	0.0035	0.0050
	(0.0198)	(0.0072)	(0.0100)	(0.0133)	(0.0027)	(0.0036)
$\theta_0 = (0.001, 0, 0.01)$	$\hat{\alpha}$	$\hat{\beta}$	$\hat{\sigma}$	$\hat{\alpha}$	$\hat{\beta}$	$\hat{\sigma}$
Empirical mean	0.019	-0.0055	0.0147	0.0331	-0.0004	0.017
	(0.0183)	(0.0063)	(0.0089)	(0.0325)	(0.0106)	(0.0096)
Empirical std.dev.	0.0849	0.0339	0.033	0.1589	0.059	0.0389
	(0.0849)	(0.0337)	(0.0321)	(0.1588)	(0.0581)	(0.0383)
Empirical MSE	0.0075	0.0012	0.0011	0.0263	0.0035	0.0016
	(0.0072)	(0.0011)	(0.0010)	(0.0252)	(0.0035)	(0.0015)
$\theta_0 = (0.1, 0.1, 0.1)$	$\hat{\alpha}$	$\hat{\beta}$	$\hat{\sigma}$	$\hat{\alpha}$	$\hat{\beta}$	$\hat{\sigma}$
Empirical mean	0.1008	0.1011	0.1222	0.0961	0.1024	0.1159
	(0.0211)	(0.0151)	(0.0268)	(0.0146)	(0.0095)	(0.0204)
Empirical std.dev.	0.0520	0.0417	0.0430	0.0275	0.0283	0.0350
	(0.0475)	(0.0389)	(0.0402)	(0.0235)	(0.0268)	(0.0325)
Empirical MSE	0.0027	0.0017	0.0023	0.0008	0.00081	0.0015
	(0.0027)	(0.0017)	(0.0016)	(0.0007)	(0.0008)	(0.0011)
$\theta_0 = (0.1, -0.1, 0.1)$	$\hat{\alpha}$	$\hat{\beta}$	$\hat{\sigma}$	$\hat{\alpha}$	$\hat{\beta}$	$\hat{\sigma}$
Empirical mean	0.1116	-0.1205	0.165	0.1002	-0.1106	0.1355
	(0.0433)	(0.0488)	(0.0741)	(0.0321)	(0.0377)	(0.0476)
Empirical std.dev.	0.0807	0.0867	0.1109	0.0441	0.0650	0.0697
	(0.0690)	(0.0744)	(0.1050)	(0.0301)	(0.0540)	(0.0620)
Empirical MSE	0.0066	0.0079	0.0165	0.0019	0.0043	0.0061
	(0.0058)	(0.0063)	(0.0111)	(0.0019)	(0.0036)	(0.0040)

Table 6.1 *Estimates and the empirical means of the absolute values of the estimated biases (in the brackets).*

	$\widehat{\alpha}$ (T=512)	$\widehat{\beta}$ (T=512)	$\widehat{\sigma}$ (T=512)	$\widehat{\alpha}$ (T=1024)	$\widehat{\beta}$ (T=1024)	$\widehat{\sigma}$ (T=1024)
$\theta_0 = (0.3, 0.01, 0.1)$						
Empirical mean	0.3167 (0.0806)	-0.0136 (0.0707)	0.1713 (0.0808)	0.3022 (0.0487)	0.0021 (0.0587)	0.1326 (0.0434)
Empirical std.dev.	0.1272 (0.0995)	0.1220 (0.1020)	0.1343 (0.1287)	0.0816 (0.0653)	0.1018 (0.0834)	0.0786 (0.0732)
Empirical MSE	0.0164 (0.0140)	0.0154 (0.0126)	0.0231 (0.0166)	0.0067 (0.0064)	0.0104 (0.0095)	0.0072 (0.0055)
$\theta_0 = (0.3, -0.01, 0.1)$						
Empirical mean	0.3191 (0.0685)	-0.0444 (0.0658)	0.1767 (0.0837)	0.296 (0.0574)	-0.0174 (0.0687)	0.1412 (0.0551)
Empirical std.dev.	0.1186 (0.0984)	0.1220 (0.1083)	0.1297 (0.1252)	0.0876 (0.0660)	0.1167 (0.0944)	0.0909 (0.0831)
Empirical MSE	0.0144 (0.0121)	0.0161 (0.0127)	0.0227 (0.0157)	0.0077 (0.0072)	0.0137 (0.0127)	0.0100 (0.0071)
$\theta_0 = (0.5, 0.45, 0.1)$						
Empirical mean	0.5349 (0.0628)	0.4217 (0.0462)	0.1475 (0.0535)	0.4982 (0.0185)	0.442 (0.0283)	0.1210 (0.0267)
Empirical std.dev.	0.1473 (0.1376)	0.0896 (0.0817)	0.1036 (0.1006)	0.0489 (0.0453)	0.0467 (0.0379)	0.0442 (0.0409)
Empirical MSE	0.0229 (0.0197)	0.0088 (0.0070)	0.0130 (0.0101)	0.0024 (0.0023)	0.0022 (0.0018)	0.0024 (0.0017)
$\theta_0 = (0.5, -0.45, 0.1)$						
Empirical mean	0.4922 (0.0483)	-0.4432 (0.0604)	0.1647 (0.1038)	0.4901 (0.0491)	-0.4465 (0.0623)	0.1619 (0.0857)
Empirical std.dev.	0.0644 (0.0430)	0.0915 (0.0687)	0.1329 (0.1050)	0.0615 (0.0380)	0.0873 (0.0610)	0.1002 (0.0805)
Empirical MSE	0.0042 (0.0035)	0.0084 (0.0076)	0.0218 (0.0125)	0.0039 (0.0030)	0.0076 (0.0072)	0.0139 (0.0070)

Table 6.2 *Estimates and the empirical means of the absolute values of the estimated biases (in the brackets).*

	T = 512			T = 1024		
$\theta_0 = (0.2, 0.3, 1)$	$\widehat{\alpha}$	$\widehat{\beta}$	$\widehat{\sigma}$	$\widehat{\alpha}$	$\widehat{\beta}$	$\widehat{\sigma}$
Empirical mean	0.2042 (0.0506)	0.2992 (0.0640)	1.0136 (0.0683)	0.2000 (0.0458)	0.2906 (0.0619)	1.0179 (0.0502)
Empirical std.dev.	0.0702 (0.0486)	0.0926 (0.0666)	0.0873 (0.0557)	0.0648 (0.0456)	0.0831 (0.0559)	0.0651 (0.045)
Empirical MSE	0.0049 (0.0045)	0.0086 (0.0084)	0.0078 (0.0061)	0.0042 (0.0042)	0.0700 (0.0058)	0.0044 (0.0031)
$\theta_0 = (0.2, -0.3, 1)$	$\widehat{\alpha}$	$\widehat{\beta}$	$\widehat{\sigma}$	$\widehat{\alpha}$	$\widehat{\beta}$	$\widehat{\sigma}$
Empirical mean	0.191 (0.0663)	-0.2717 (0.0643)	1.0171 (0.0672)	0.181 (0.0616)	-0.2759 (0.0474)	1.0363 (0.0689)
Empirical std.dev.	0.1025 (0.0784)	0.0799 (0.0549)	0.0871 (0.0576)	0.0709 (0.0396)	0.0577 (0.0406)	0.0836 (0.0593)
Empirical MSE	0.0106 (0.0094)	0.0072 (0.0043)	0.0079 (0.0058)	0.0054 (0.0034)	0.0039 (0.0022)	0.0083 (0.0046)
$\theta_0 = (0.4, 0.2, 10)$	$\widehat{\alpha}$	$\widehat{\beta}$	$\widehat{\sigma}$	$\widehat{\alpha}$	$\widehat{\beta}$	$\widehat{\sigma}$
Empirical mean	0.4083 (0.0401)	0.1812 (0.0438)	9.9999 (0.0449)	0.3962 (0.0305)	0.1927 (0.0263)	10.0025 (0.0315)
Empirical std.dev.	0.0681 (0.0555)	0.0587 (0.0431)	0.0901 (0.0779)	0.0405 (0.0267)	0.0332 (0.0215)	0.0479 (0.0361)
Empirical MSE	0.0047 (0.0041)	0.0038 (0.0025)	0.0081 (0.0081)	0.0016 (0.0014)	0.0012 (0.0008)	0.002 (0.0021)
$\theta_0 = (0.4, -0.2, 10)$	$\widehat{\alpha}$	$\widehat{\beta}$	$\widehat{\sigma}$	$\widehat{\alpha}$	$\widehat{\beta}$	$\widehat{\sigma}$
Empirical mean	0.4008 (0.0586)	-0.2142 (0.0381)	10.0095 (0.0789)	0.3893 (0.0513)	-0.2134 (0.0344)	10.0276 (0.0785)
Empirical std.dev.	0.0812 (0.0559)	0.0474 (0.0314)	0.1232 (0.0948)	0.0630 (0.0377)	0.0404 (0.0249)	0.1083 (0.0793)
Empirical MSE	0.0066 (0.0065)	0.0024 (0.0016)	0.0153 (0.0138)	0.0041 (0.0031)	0.0018 (0.0011)	0.0125 (0.0089)

Table 6.3 *Estimates and the empirical means of the absolute values of the estimated biases (in the brackets).*

	$T = 512$			$T = 1024$		
$\theta_0 = (0.1, 0.2, 0.01)$	$\hat{\alpha}$	$\hat{\beta}$	$\hat{\sigma}$	$\hat{\alpha}$	$\hat{\beta}$	$\hat{\sigma}$
Empirical mean	0.2674 (0.1764)	0.1718 (0.046)	0.3007 (0.2936)	0.2937 (0.2018)	0.1907 (0.046)	0.5467 (0.5402)
Empirical std.dev.	0.7007 (0.6984)	0.1100 (0.1038)	1.8803 (1.8798)	0.6492 (0.6467)	0.0965 (0.0852)	2.3606 (2.3598)
Empirical MSE	0.5190 (0.4878)	0.0129 (0.0111)	3.6200 (3.5336)	0.4590 (0.4183)	0.0094 (0.0086)	5.8605 (5.5687)
$\theta_0 = (0.1, -0.2, 0.01)$	$\hat{\alpha}$	$\hat{\beta}$	$\hat{\sigma}$	$\hat{\alpha}$	$\hat{\beta}$	$\hat{\sigma}$
Empirical mean	0.1487 (0.0612)	-0.2183 (0.0537)	0.1012 (0.0947)	0.1247 (0.0496)	-0.2226 (0.0411)	0.0772 (0.0712)
Empirical std.dev.	0.1152 (0.109)	0.0981 (0.0839)	0.1513 (0.1491)	0.0901 (0.0790)	0.0760 (0.0677)	0.1173 (0.1149)
Empirical MSE	0.0156 (0.0120)	0.0098 (0.0083)	0.0312 (0.0222)	0.0087 (0.0068)	0.0063 (0.0049)	0.0183 (0.0132)
$\theta_0 = (1, 0.2, 0.01)$	$\hat{\alpha}$	$\hat{\beta}$	$\hat{\sigma}$	$\hat{\alpha}$	$\hat{\beta}$	$\hat{\sigma}$
Empirical mean	0.9971 (0.0375)	0.1532 (0.0955)	0.1219 (0.113)	0.9967 (0.0258)	0.1762 (0.0758)	0.0937 (0.0851)
Empirical std.dev.	0.0573 (0.0433)	0.1676 (0.1452)	0.1804 (0.1797)	0.0353 (0.0242)	0.1360 (0.1152)	0.1319 (0.1309)
Empirical MSE	0.0033 (0.0031)	0.0302 (0.0234)	0.0451 (0.0323)	0.0012 (0.0011)	0.0191 (0.0160)	0.0244 (0.017)
$\theta_0 = (1, -0.2, 0.01)$	$\hat{\alpha}$	$\hat{\beta}$	$\hat{\sigma}$	$\hat{\alpha}$	$\hat{\beta}$	$\hat{\sigma}$
Empirical mean	0.9899 (0.0339)	-0.1978 (0.0498)	0.1223 (0.1162)	0.9947 (0.0266)	-0.2086 (0.0425)	0.1052 (0.097)
Empirical std.dev.	0.0512 (0.0395)	0.0878 (0.0721)	0.1811 (0.1786)	0.0368 (0.0259)	0.0788 (0.0668)	0.1409 (0.1396)
Empirical MSE	0.0027 (0.0021)	0.0077 (0.0075)	0.0454 (0.0319)	0.0014 (0.0011)	0.0063 (0.0056)	0.0289 (0.0195)

Table 6.4 *Estimates and the empirical means of the absolute values of the estimated biases (in the brackets).*

choice of β values. These results in Table 6.2 show that there is some MSE distortion in the case of $T = 512$ for β_0 when β_0 is as small as either 0.01 or -0.01. When T increases to 1024, the MSEs become stable. For the case of $\beta_0 = 0.45$ or -0.45, the MSEs look quite stable and very small.

In Tables 6.3 and 6.4, both relatively large and relatively small values for σ_0 have also been considered. When the volatility σ_0 of the FSV model is as small as 0.1, Table 6.2 shows that the MSEs are both stable and quite small. For the case of $\sigma_0 = 10$ in the second section of Table 6.3, the MSEs for the estimates of all the components of θ_0 are quite stable and very small, particularly when $T = 1024$.

Since empirical financial evidence also suggests that very small volatility parameter values make precise estimation quite difficult, we consider four cases in Table 6.4 that the volatility parameter value is as small as $\sigma_0 = 0.01$. These results show that the MSEs become quite reasonable when the sample size is as medium as $T = 1024$. It is also observed from Table 6.4 that the change of the drift parameter value from $\alpha_0 = 0.1$ to $\alpha_0 = 1$ does not affect the MSEs significantly.

In summary, the MSEs in Tables 6.1–6.4 demonstrate that both the proposed estimation procedure and the asymptotic convergence established in Theorem 6.4 work well numerically. In addition, both the proposed theory and estimation procedure have been applied to several market indexes in the following section.

6.4.4 Applications to market indexes

Market indexes are a guideline of investor confidence and are inter–related to the performance of local and global economies. Investments on market indexes such as the Dow Jones, S&P 500, FTSE 100, etc, are common practice. In this section, we apply model (6.32) to model the stochastic volatility and then the estimation procedure to assess both the memory and volatility properties of such indexes. The first part of this section provides the explicit expressions of both the mean and the variance functions of stochastic volatility process $V(t)$. The second part of this section describes briefly these indexes. Some empirical results are given in the last part of this section. The discussion of this section is based on the paper by Casas and Gao (2006).

Estimation of mean and variance

In addition to estimating the three parameters α, β and σ involved in model (6.32), it is also interesting to estimate both the mean μ_V and the standard deviation σ_V of the stochastic volatility process $V(t)$.

Since the stationary version $V(t) = e^{X(t)}$ in model (6.32) is considered, it can be shown that both the mean and variance functions of $V(t)$ may be expressed as follows:

$$
\begin{aligned}
\mu_V &= \mu_V(\theta) = \exp(\sigma_X^2/2) \quad \text{and} \\
\sigma_V^2 &= \sigma_V^2(\theta) = \left(\exp(\sigma_X^2 - 1)\right)\exp(\sigma_X^2),
\end{aligned}
\tag{6.34}
$$

where

$$
\sigma_X^2 = \sigma_X^2(\theta) = \gamma_X(0) = \frac{\sigma^2 \pi}{\Gamma^2(1+\beta)\alpha^{1+2\beta}\cos(\beta\pi)}.
\tag{6.35}
$$

The mean $\mu_V(\theta)$ and the standard deviation $\sigma_V(\theta)$ are then estimated by

$$
\widehat{\mu}_V = \mu_V(\widehat{\theta}_T) \quad \text{and} \quad \widehat{\sigma}_V = \sigma_V(\widehat{\theta}_T)
\tag{6.36}
$$

with $\widehat{\theta}_T$ being defined before. For the market indexes, the corresponding estimates are given in Tables 6.5 and 6.6 below. Some detailed descriptions about the real data are first given below.

Real data

The data sets chosen for our empirical study are: a) two major American indexes: Dow Jones Industrial Average and S&P 500; b) three of the major European indexes: CAC 40, DAX 30 and FTSE 100; and c) the major Asian index, the NIKKEI 225. A vast amount of information about market indexes is available on the Internet. Wikipedia and InvestorWords.com give easy access to informative glossaries of the stock market indexes.

Dow Jones Industrial Average: The DJIA started in July 1884 with the main railway companies of the time. Today the Dow is calculated as the price–weighted average of 30 blue chip stocks from each important stock sector in the market (except transportation and utilities): chemical, steel, tobacco, sugar, electrics, motors, retail, etc. The section under study in this section is from October 1, 1928 to July 29, 2005 and can be seen in Figure 6.3. More information can be found at the Dow Jones, Dow Jones Indexes corporate sites and the Wall Street Journal amongst others.

S&P 500: Standard & Poor's 500 is a market–value weighted price of 500 stocks: some from the New York Stock Exchange and, since 1973,

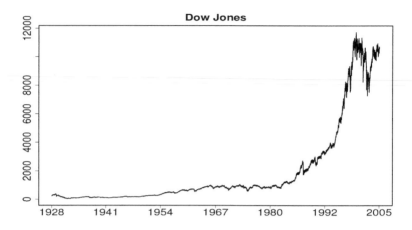

Figure 6.3 *Dow Jones Industrial Average.*

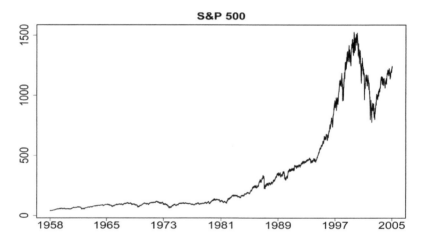

Figure 6.4 *S&P 500.*

some from the NASDAQ stock market. The choice of stock prices aim
to achieve a common distribution between the grouping price and the
distribution of the total New York Stock Exchange. The S&P 500 Index
Committee establishes the guidelines for addition and deletion of prices
into the index. It was created in 1957, but its values have already been
extrapolated since the beginning of 1928. It represents 70% of the U.S.
equity market. The values in Figure 6.4 are from January 2, 1958 to July
29, 2005.

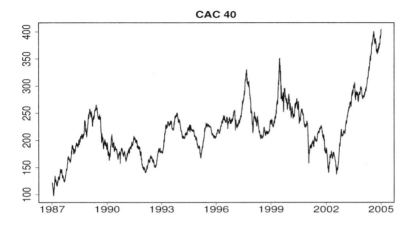

Figure 6.5 *French CAC 40.*

CAC 40: The CAC 40 is the French Stock Market Index which was set up on December 30, 1987. It is a market–weighted average of the 40 most significant values out of the top 100 market capitalisations on the Paris Bourse. An interesting feature of this market is that the 45% of the shares belong to foreign investors, therefore fluctuations of the CAC 40 can be an indicator of international economy performance. Figure 6.5 plots the values from December 30, 1987 to July 29, 2005.

DAX 30: Trading on the Frankfurt Stock Exchange, the DAX 30 is a price–weighted index of the 30 top German companies in terms of book volume and market capitalisation. Figure 6.6 shows the index from its beginning on December 31, 1964 to July 29, 2005. More information can be found in the web page of Gruppe Deutsche Börse.

FTSE 100: The Financial Times Stock Exchange 100 is a guideline of the performance of the British economy and one of the most important indexes in Europe. It is a market–weighted index of the largest 100 stocks in the London Stock Exchange. The index from January 31, 1978 to July 29, 2005 is shown in Figure 6.7. The FTSE homepage contains more detailed information.

NIKKEI 225 Average: This is a price–weighted average of 225 stocks of the Tokyo Stock Exchange from industry and technology. It has been calculated daily since 1971. The performance of this index is different from the American indexes. The NIKKEI increased fiercely from the 70's until the end of the 80's, experiencing a drop of over 30,000 within

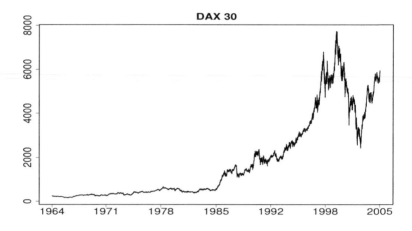

Figure 6.6 *German DAX 30.*

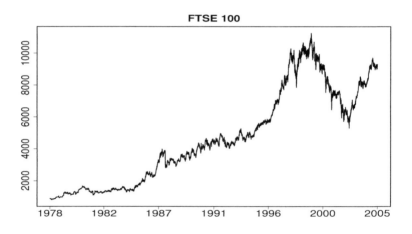

Figure 6.7 *Financial Times Stock Exchange 100.*

months and the descending trend has continued until today. Figure 6.8 shows the index values since January 4, 1984 to July 29, 2005.

Empirical results

We have applied the proposed estimation procedure to real financial data. The results in Table 6.5 contain the estimates of α, β, σ, μ_V, and σ_V.

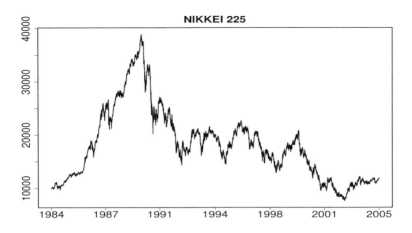

Figure 6.8 *Japanese NIKKEI 225 Average.*

It is not surprising that the parameter estimates of the DJIA and S&P
500 are comparable as Table 6.5 shows. A quick look at Figures 6.3 and
6.4 shows that the two American indexes behave similarly. In both cases
$\widehat{\beta}$ is about 1%, meaning that the volatility of these indexes displays
a certain degree of LRD. This is expected from the studies of Ding,
Granger and Engle (1993) who showed that the square of the returns
of the S&P 500 displays LRD. The two indexes have $\widehat{\sigma}$ of 1% but differ
from each other in the value of $\widehat{\alpha}$. We believe that this is due to the
difference in periods of the data sets, as the DJIA was studied from 1928
and the S&P 500 from 1958. This difference results in a greater standard
deviation of the volatility process for the DJIA, 56%, in comparison with
a standard deviation of 44% for the S&P 500.

Results of Table 6.5 for the European indexes suggest that the volatility
of the French CAC 40 and the British FTSE 100 display IRD or antipersistence. In contrast, the German DAX 30 does not seem to have any
memory of the past. The mean $\widehat{\mu}_V$ is 1.06 for the DAX 30 and 1.07 for
the CAC 40 and FTSE 100. Thus, the expected value of the volatility
is roughly the same for the three European indexes. In addition, the
German index is less risky because the standard deviation of the volatility process is 37% in comparison with 43% in the other two European
indexes.

The estimates of α and σ for the NIKKEI 225 are larger than those of
the American and European indexes. In fact, $\widehat{\alpha}$ of the NIKKEI is three
times larger than $\widehat{\alpha}$ of the Dow Jones, and $\widehat{\sigma}$ of the NIKKEI is nine
times larger than $\widehat{\sigma}$ of the Dow Jones. Thus, for the same value of $\widehat{\beta}$,

$\widehat{\sigma}_V$ of the NIKKEI is expected to be roughly 50 times larger than $\widehat{\sigma}_V$ of the Dow Jones. However, as we can see in Table 6.5, the NIKKEI has a standard deviation of 53% and the Dow Jones has a dispersion of 56%. This shows that the LRD parameter influences the standard deviation of the volatility greatly. A series with a positive β will tend to have a larger standard deviation in the volatility than a series with antipersistence.

In an attempt to understand the evolution of the volatility through time and in particular the behaviour of the LRD parameter, three sections of the DJIA are considered separately. The results are given in Table 6.6. The β parameter does not change through time but α and σ do. The first section of the data from 1928 to 1968 registers the larger values for the mean and standard deviation of the volatility process. This is a consequence of the Great Depression of 1929 and the following economic recession that severely hit the U.S. economy. It was not until well after World War II that indicators of industrial production such as the GDP and share prices reached the values they had prior to 1929. Sections from 1928 to 1988 and from 1928 to 2005 have experienced a decrease of the mean and standard deviation of the volatility process. The last 35 years have not passed without financial shocks such as the consequence of the oil crisis of 1973, the Black Monday of 1987 and two Gulf Wars. However, the impacts of these events have not affected the DJIA as strongly as the Great Depression whose effects are easing off with time. We may say that nowadays the DJIA is experiencing a time of stability.

In summary, the volatility of each of the seven major worldwide stock indexes has been estimated. The DJIA and the S&P 500 display LRD, so statistical patterns may be repeated at different time scales producing arbitrage opportunities. The volatility of the DJIA and the NIKKEI 225 have the largest dispersion, making them riskier than the other indexes. The European indexes have the lowest standard deviation in the volatility process, in particular the German index. The three European indexes and the Japanese index display IRD.

6.5 Technical notes

6.5.1 Proof of Theorem 6.6

This appendix provides only an outline of the proof of Theorem 6.6, since some technical details are quite standard but tedious in this kind of proof and therefore omitted here. To prove Theorem 6.6, we need to introduce the following lemmas.

Lemma 6.1. *Suppose that Assumptions 6.12 and 6.13 hold. Then as* $T \to \infty$

$$W_T(\theta) \to W(\theta) = \frac{1}{4\pi} \int_{-\pi}^{\pi} \left\{ \log(f_Z(\omega, \theta)) + \frac{f_Z(\omega, \theta_0)}{f_Z(\omega, \theta)} \right\} d\omega. \quad (6.37)$$

PROOF: In order to prove (6.37), it suffices to show that for every continuous function $w(\omega, \theta)$

$$\int_{-\pi}^{\pi} I_T(\omega) w(\omega, \theta) \, d\omega \to \int_{-\pi}^{\pi} f_Z(\omega, \theta_0) w(\omega, \theta) \, d\omega \quad (6.38)$$

with probability one as $T \to \infty$.

In view of the expression of $Z_t = \mu + 2X_t + e_t$ in (6.25) as well as Lemma 1 of Hannan (1973), the proof of Fox and Taqqu (1986) remains valid for the case where $\{Z_t\}$ is non–Gaussian. This is mainly because of the following three reasons.

The first reason is that Lemma 1 of Hannan (1973) is applicable to non–Gaussian time series and thus to our time series $\{Z_t\}$. The second reason is that the proof of Lemma 1 of Fox and Taqqu (1986) remains valid for $Z_t = \mu + 2X_t + e_t$, in which $\{X_t\}$ is the Gaussian time series with LRD but $\{e_t\}$ is a sequence i.i.d. random errors. The third reason is that Assumption 6.12(ii) guarantees that $\{X_t\}$ admits a backward expansion of the form (see Fox and Taqqu 1986, p. 520)

$$X_t = \sum_{s=0}^{\infty} b_s u_{t-s}, \quad (6.39)$$

where $\{b_s\}$ is a sequence of suitable real numbers such that $\{X_t\}$ is a Gaussian process with its spectral density function being given by $f_X(\omega, \theta)$, and $\{u_s\}$ is a sequence of independent and Normally distributed random variables with $E[u_s] = 0$ and $E[u_s^2] = 1$.

Lemma 6.2. *Suppose that Assumptions 6.12 and 6.13 hold. Then as* $T \to \infty$

$$\widehat{\theta}_T - \theta_0 \to 0 \quad \text{with probability one.} \quad (6.40)$$

PROOF: Lemma 6.1 implies that the following holds with probability one for $\theta \neq \theta_0$,

$$W_T(\theta) - W_T(\theta_0) \to L(\theta, \theta_0) \quad (6.41)$$

as $T \to \infty$, where

$$L(\theta, \theta_0) = \frac{1}{4\pi} \int_{-\pi}^{\pi} \left\{ \frac{f_Z(\omega, \theta_0)}{f_Z(\omega, \theta)} - 1 - \log\left(\frac{f_Z(\omega, \theta_0)}{f_Z(\omega, \theta)} \right) \right\} d\omega > 0.$$

Index	$\widehat{\alpha}$	$\widehat{\beta}$	$\widehat{\sigma}$	$\widehat{\mu}_V$	$\widehat{\sigma}_V$
DJIA	0.0029	0.0109	0.0127	1.1181	0.5592
S&P 500	0.0047	0.0108	0.0143	1.0808	0.4432
CAC 40	0.0047	-0.2778	0.0667	1.0764	0.4288
DAX 30	0.0054	0.0059	0.0137	1.0602	0.3734
FTSE 100	0.0013	-0.3112	0.0606	1.0760	0.4274
NIKKEI 225	0.0094	-0.2752	0.0912	1.1088	0.5312

Table 6.5 *Estimates of market indexes volatility parameters.*

Dow Jones Industrial Average	$\widehat{\alpha}$	$\widehat{\beta}$	$\widehat{\sigma}$	$\widehat{\mu}_V$	$\widehat{\sigma}_V$
1 Oct. 1928 – 23 Sep. 1968	0.0035	0.0108	0.0160	1.1405	0.6255
1 Oct. 1928 – 27 Jul. 1988	0.0033	0.0109	0.0138	1.1096	0.5336
1 Oct. 1928 – 29 Jul. 2005	0.0029	0.0109	0.0127	1.1056	0.5214

Table 6.6 *Evolution of the volatility through time.*

Thus, for any given $\epsilon > 0$

$$\lim_{T \to \infty} \inf \inf_{||\theta - \theta_0|| \geq \epsilon} (W_T(\theta) - W_T(\theta_0)) > 0 \qquad (6.42)$$

with probability one. The proof of $\widehat{\theta}_T \to \theta_0$ with probability one follows from Assumption 6.12(iii). Thus, the proof of Lemma 6.2 is finished.

PROOF OF THEOREM 6.6(i). The first part of Theorem 6.6 has already been proved in Lemma 6.2.

PROOF OF THEOREM 6.6(ii): The proof of the second part of Theorem 6.6 is standard. Note that Assumption 6.12 implies that Condition (A1) of Heyde and Gay (1993) is satisfied because of (6.39) and the expression of $Z_t = \mu + 4X_t + e_t$. Assumption 6.13 implies that conditions (A2) and (A3) of Heyde and Gay (1993) are satisfied for $f_X(\omega, \theta)$ and then $f_Z(\omega, \theta)$. Thus, by applying the mean–value theorem and Theorem 1(ii) of Heyde and Gay (1993), the proof is completed.

6.6 Bibliographical notes

In the field of time series with long–range dependence (long memory), the literature is huge. Recent survey papers and monographs include Beran (1994), Robinson (1994), Baillie and King (1996), Anh and Heyde (1999), and Robinson (2003) for general developments on long-range dependence in econometrics and statistics. Other related references on parameter estimation include Mandelbrot and Van Ness (1968), Geweke and Porter–Hudak (1983), Fox and Taqqu (1986), Künsch (1986), Dahlhaus (1989), Heyde and Gay (1993), Hurvich and Beltrao (1993), Viano, Deniau and Oppenheim (1994, 1995), Ding and Granger (1996), Hurvich, Deo and Brodsky (1998), Gao *et al.* (2001), Gao, Anh and Heyde (2002), Sun and Phillips (2003), Gao (2004), Lieberman and Phillips (2004, 2005), Robinson (2005), and Shimotsu and Phillips (2005, 2006). Meanwhile, existing studies in the field of stochastic volatility with long–range dependence include Taylor (1986, 1994), Robinson and Zaffaroni (1998), Robinson (2001), Broto and Ruiz (2004), Berkes *et al.* (2006), and Casas and Gao (2006).

In the statistics literature, there are a number of papers discussing nonparametric and semiparametric estimation of time series regression with long–range dependence. Existing studies include Hall and Hart (1990), Cheng and Robinson (1994), Hall, Lahiri and Polzehl (1995), Robinson (1997), Beran and Ghosh (1998), Koul and Stute (1998), Anh *et al.* (1999), Gao and Anh (1999), Beran and Feng (2002), and Beran, Ghosh and Sibbertsen (2003). By contrast, there is little research about model specification of nonlinear time series with long–range dependence. To the best of our knowledge, the only paper available is Gao and Wang (2006).

CHAPTER 7

Appendix

The following lemmas and theorems are of general interest and can be used for other nonlinear estimation and testing problems associated with the α–mixing condition. Lemma A.2 and Theorem A.1 below extend some corresponding results of Hjellvik, Yao and Tjøstheim (1998) and Fan and Li (1999) for the β–mixing case.

Theorem A.1 improves existing results in the field, such as Gao and Anh (2000), and Lemma A.1 of Gao and King (2004). Theorem A.2 is a novel result of this kind for the α–mixing time series case and has some useful applications in deriving asymptotic Edgeworth expansions for kernel–based test statistics.

7.1 Technical lemmas

Lemma A.1. *Suppose that M_m^n are the σ-fields generated by a stationary α-mixing process ξ_i with the mixing coefficient $\alpha(i)$. For some positive integers m let $\eta_i \in M_{s_i}^{t_i}$ where $s_1 < t_1 < s_2 < t_2 < \cdots < t_m$ and suppose $t_i - s_i > \tau$ for all i. Assume further that*

$$||\eta_i||_{p_i}^{p_i} = E|\eta_i|^{p_i} < \infty,$$

for some $p_i > 1$ for which $Q = \sum_{i=1}^{l} \frac{1}{p_i} < 1$. Then

$$\left| E\left[\prod_{i=1}^{l} \eta_i \right] - \prod_{i=1}^{l} E[\eta_i] \right| \le 10(l-1)\alpha(\tau)^{(1-Q)} \prod_{i=1}^{l} ||\eta_i||_{p_i}.$$

PROOF: See Roussas and Ionnides (1987).

Lemma A.2. (i) *Let $\psi(\cdot, \cdot, \cdot)$ be a symmetric Borel function defined on $R^r \times R^r \times R^r$. Let the process ξ_i be a r-dimensional strictly stationary and strong mixing (α–mixing) stochastic process. Assume that for any* *fixed* $x, y \in R^r$, $E[\psi(\xi_1, x, y)] = 0$. *Then*

$$E\left\{ \sum_{1 \leq i < j < k \leq T} \psi(\xi_i, \xi_j, \xi_k) \right\}^2 \leq CT^3 M^{\frac{1}{1+\delta}},$$

where $0 < \delta < 1$ *is a small constant,* $C > 0$ *is a constant independent of* T *and the function* ψ, $M = \max\{M_1, M_2, M_3, M_4\}$, *and*

$$M_1 = \max_{1 < i < j \leq T} \max \left\{ E|\psi(\xi_1, \xi_i, \xi_j)|^{2(1+\delta)} \right\},$$

$$M_2 = \max_{1 < i < j \leq T} \max \left\{ \int |\psi(\xi_1, \xi_i, \xi_j)|^{2(1+\delta)} dP(\xi_1) dP(\xi_i, \xi_j) \right\},$$

$$M_3 = \max_{1 < i < j \leq T} \max \left\{ \int |\psi(\xi_1, \xi_i, \xi_j)|^{2(1+\delta)} dP(\xi_j) dP(\xi_1, \xi_i) \right\},$$

$$M_4 = \max_{1 < i < j \leq T} \max \left\{ \int |\psi(\xi_1, \xi_i, \xi_j)|^{2(1+\delta)} dP(\xi_1) dP(\xi_i) dP(\xi_j) \right\}.$$

(ii) *Let* $\phi(\cdot, \cdot)$ *be a symmetric Borel function defined on* $R^r \times R^r$. *Let the process* ξ_i *be defined as in (i). Assume that for any fixed* $x \in R^r$, $E[\phi(\xi_1, x)] = 0$. *Then*

$$E\left\{ \sum_{1 \leq i < j \leq T} \phi(\xi_i, \xi_j) \right\}^2 \leq CT^2 M_5^{\frac{1}{1+\delta}},$$

where $\delta > 0$ *is a constant,* $C > 0$ *is a constant independent of* T *and the function* ϕ, *and*

$$M_5 = \max_{1 < i < j \leq T} \max \left\{ E|\phi(\xi_1, \xi_i)|^{2(1+\delta)}, \int |\phi(\xi_1, \xi_i)|^{2(1+\delta)} dP(\xi_1) dP(\xi_i) \right\}.$$

PROOF: As the proof of (ii) is similar to that of (i), one proves only (i). Let $i_1, \ldots, i_6$ be distinct integers and $1 \leq i_j \leq T$, let $1 \leq k_1 < \cdots < k_6 \leq T$ be the permutation of $i_1, \ldots, i_6$ in ascending order and let d_c be the c-th largest difference among $k_{j+1} - k_j$, $j = 1, \cdots, 5$. Let

$$H(k_1, \cdots, k_6) = \psi(\xi_{i_1}, \xi_{i_2}, \xi_{i_3}) \psi(\xi_{i_4}, \xi_{i_5}, \xi_{i_6}).$$

By Lemma A.1 (with $\eta_1 = \psi(\xi_{i_1}, \xi_{i_2}, \xi_{i_3})$, $\eta_2 = \psi(\xi_{i_4}, \xi_{i_5}, \xi_{i_6})$, $l = 2$, $p_i = 2(1+\delta)$ and $Q = \frac{1}{1+\delta}$),

$$|E[H(k_1, \cdots, k_6)]| \leq \begin{cases} 10 M^{\frac{1}{1+\delta}} \alpha^{\frac{\delta}{1+\delta}} (k_6 - k_5) & \text{if } k_6 - k_5 = d_1 \\ 10 M^{\frac{1}{1+\delta}} \alpha^{\frac{\delta}{1+\delta}} (k_2 - k_1) & \text{if } k_2 - k_1 = d_1. \end{cases}$$

Thus,

$$\sum_{\substack{1 \le k_1 < \cdots < k_6 \le T \\ k_2 - k_1 = d_1}} |E[H(k_1, \cdots, k_6)]|$$

$$\le \sum_{k_1=1}^{T-5} \sum_{k_2=k_1+\max_{j\ge 3}\{k_j-k_{j-1}\}} \sum_{k_3=k_2+1}^{T-3} \cdots \sum_{k_6=k_5+1}^{T}$$

$$\times \left\{ 10M^{\frac{1}{1+\delta}} \alpha^{\frac{\delta}{1+\delta}} (k_2 - k_1) \right\}$$

$$\le 10M^{\frac{1}{1+\delta}} \sum_{k_1=1}^{T-5} \sum_{k_2=k_1+1}^{T-4} (k_2 - k_1)^2 \alpha^{\frac{\delta}{1+\delta}}(k_2 - k_1)$$

$$\le 10TM^{\frac{1}{1+\delta}} \sum_{k=1}^{T} k^4 \alpha^{\frac{\delta}{1+\delta}}(k) \le CTM^{\frac{1}{1+\delta}}.$$

Similarly,

$$\sum_{\substack{1 \le k_1 < \cdots < k_6 \le T \\ k_2 - k_1 = d_1}} |E[H(k_1, \cdots, k_6)]| \le CTM^{\frac{1}{1+\delta}}.$$

Analogously, it can be shown in a similar way that

$$\sum_{\substack{1 \le k_1 < \cdots < k_6 \le T \\ k_6 - k_5 = d_2 \text{ or } k_2 - k_1 = d_2}} |E[H(k_1, \cdots, k_6)]| \le CT^2 M^{\frac{1}{1+\delta}},$$

$$\sum_{\substack{1 \le k_1 < \cdots < k_6 \le T \\ k_6 - k_5 = d_3 \text{ or } k_2 - k_1 = d_3}} |E[H(k_1, \cdots, k_6)]| \le CTM^{\frac{1}{1+\delta}}.$$

On the other hand, if $\{k_6 - k_5, k_2 - k_1\} = \{d_4, d_5\}$, by using Lemma A.1 three times we have the inequality

$$|E[H(k_1, \cdots, k_6)]| \le 10M^{\frac{1}{1+\delta}} \sum_{i=1}^{3} \alpha^{\frac{\delta}{1+\delta}}(d_i).$$

Hence,

$$\sum_{\substack{1 \le k_1 < \cdots < k_6 \le T \\ \{k_6 - k_5, k_2 - k_1\} = \{d_4, d_5\}}} |E[H(k_1, \cdots, k_6)]|$$

$$\leq \sum_{\substack{1 \leq k_1 < \cdots < k_6 \leq T \\ \max\{k_6 - k_5, k_2 - k_1\} \\ \leq \min_{2 \leq j \leq 4}\{k_{j+1} - k_j\}}}$$

$$\times \left\{ 10M^{\frac{1}{1+\delta}} \left[\alpha^{\frac{\delta}{1+\delta}}(k_3 - k_2) + \alpha^{\frac{\delta}{1+\delta}}(k_4 - k_3) + \alpha^{\frac{\delta}{1+\delta}}(k_5 - k_4) \right] \right\}$$

$$\leq 30M^{\frac{1}{1+\delta}} \sum_{\substack{1 \leq k_1 < \cdots < k_6 \leq T \\ \max\{k_6 - k_5, k_2 - k_1\} \leq d_3}} \alpha^{\frac{\delta}{1+\delta}}(d_3) \leq 30CT^3 M^{\frac{1}{1+\delta}}.$$

Thus, we have

$$\sum_{\substack{1 \leq i,j,k,r,s,t \leq T \\ i,j,k,r,s,t \text{ different}}} |E[\psi(\xi_i, \xi_j, \xi_k)\psi(\xi_r, \xi_s, \xi_t)]| \leq CT^3 M^{\frac{1}{1+\delta}}.$$

Similarly, we can show that

$$\sum_{\substack{1 \leq i,j,k,r,s,t \leq T \\ i,j,k,s,t \text{ different}}} |E[\psi(\xi_i, \xi_j, \xi_k)\psi(\xi_i, \xi_s, \xi_t)]| \leq CT^3 M^{\frac{1}{1+\delta}},$$

$$\sum_{\substack{1 \leq i,j,k,l \leq T \\ i,j,k,l \text{ different}}} |E[\psi(\xi_i, \xi_j, \xi_k)\psi(\xi_i, \xi_j, \xi_l)]| \leq CT^3 M^{\frac{1}{1+\delta}}.$$

Finally, it is easy to see that

$$\sum_{1 \leq i < j < k \leq T} E[\psi(\xi_i, \xi_j, \xi_k)^2] \leq T^3 \max_{1 < i < j} E[\psi(\xi_1, \xi_i, \xi_j)^2].$$

The conclusion of Lemma A.2(i) follows, and therefore the proof of Lemma A.2 is completed.

Götze, Tikhomirov and Yurchenko (2004) considered a quadratic sum of the form

$$Q_T = \sum_{j=1}^{T} a_{jj}(X_j^2 - E[X_j^2]) + \sum_{1 \leq j \neq k \leq T} a_{jk} X_j X_k, \qquad (A.1)$$

where $X_1, \cdots, X_T$ are independent and identically distributed random variables with $E[X_1] = 0$, $E[X_1^2] < \infty$ and $E[X_1^6] < \infty$. Here $\{a_{ij}\}$ is a sequence of real numbers possibly depending on T. They actually considered a more general setup than this, but for simplicity we briefly present this simplified form.

The following notational symbols are needed.

$\mathbf{A} = (a_{jk})_{j,k=1}^{T}$: $T \times T$ matrix containing all coefficients a_{jk};

$\|\mathbf{A}\| = \sum_{j,k=1}^{T} a_{jk}^{2}$;

$\mathbf{Tr}\,\mathbf{A} = \sum_{j=1}^{T} a_{jj}$: the trace of the matrix $\mathbf{A}$;

$V^{2} = \sum_{j=1}^{T} a_{jj}^{2}$ and $\mathcal{L}_{j}^{2} = \sum_{k=1}^{T} a_{jk}^{2}$, $j = 1, \cdots, T$;

$\mathbf{d}^{\tau} = (a_{11}, a_{22}, \cdots, a_{TT}) \in \mathbf{R}^{T}$: T–dimensional column vector containing all diagonal elements of the matrix $\mathbf{A}$;

$\mathbf{A_0}$ denotes the $T \times T$ matrix with elements a_{jk} if $j \neq k$ and equal to 0, if $j = k$;

$\mu_{k} = E\left[X_{1}^{k}\right]$ and $\beta_{k} = E\left[|X_{1}|^{k}\right]$ for $k = 1, \cdots, 6$;

λ_1 the maximal in absolute value, eigenvalue of the matrix $\mathbf{A}$;

$$M = \max\left(\frac{|\lambda_1|^{2}}{\|\mathbf{A}\|^{2}}, \frac{\left(\sum_{j=1}^{T} \mathcal{L}_{j}^{4}\right)^{\frac{1}{2}}}{\|\mathbf{A}\|^{2}}\right);$$

$\sigma_{*}^{2} = (\mu_4 - \mu_2^{2})V^{2} + 2\mu_2^{2}\|\mathbf{A_0}\|^{2}$ and $\kappa = \sigma_{*}^{-3}\left(\mu_3^{2}\mathbf{d}^{\tau}\mathbf{A_0}\mathbf{d} + \frac{4}{3}\mu_3^{2}\mathbf{Tr}\mathbf{A_0^{3}}\right)$.

In the above symbols the superscript τ denotes the transposition of a vector or matrix. Note that the matrix $\mathbf{A_0}$ is obtained from the matrix $\mathbf{A}$ by replacing all diagonal elements by 0.

The coefficients in the quadratic form (A.1) should satisfy some conditions:

$\mathbf{Q}$(i): $\|\mathbf{A}\| < \infty$; $\mathbf{Q}$(ii): there exists some absolute positive constant $b_1^{2} > 0$ such that

$$1 - \frac{V^{2}}{\|\mathbf{A}\|^{2}} \geq b_1^{2}.$$

Lemma A.3. *Under conditions* $\mathbf{Q}$*(i) and* $\mathbf{Q}$*(ii), we have*

$$\sup_{x}\left|P\left(\frac{Q_T}{\sigma_*} \leq x\right) - \Phi(x) + \kappa\Phi'''(x)\right| \leq Cb_1^{-4}\left(\beta_3^{2} + V\|\mathbf{A}\|^{-1}\beta_6\right)\mu_2^{-3}M,$$

where C is an absolute positive constant and where $\Phi'''(\cdot)$ denotes the third derivative of the cumulative distribution function of the standard normal.

PROOF: The proof is the same as that of Theorem 1.1 in Götze, Tikhomirov and Yurchenko (2004).

7.2 Asymptotic normality and expansions

Theorem A.1. Let ξ_t be a r-dimensional strictly stationary and $\alpha-$ mixing stochastic process with the mixing coefficient $\alpha(t) \leq C_\alpha \alpha^t$ for some $0 < C_\alpha < \infty$ and $0 < \alpha < 1$. Let $\theta(\cdot, \cdot)$ be a symmetric Borel function defined on $\mathbf{R}^r \times \mathbf{R}^r$. Assume that for any fixed $x, y \in \mathbf{R}^r$, $E[\theta(\xi_1, y)] = E[\theta(x, \xi_1)] = 0$. Let

$$\theta_{st} = \theta(\xi_s, \xi_t) \quad and \quad \sigma_T^2 = \sum_{1 \leq s < t \leq T} \mathrm{var}[\theta_{st}].$$

For some small constant $0 < \delta < 1$, let

$$M_{T11} = \max_{1 \leq i < j < k \leq T} \max \left\{ E|\theta_{ik}\theta_{jk}|^{1+\delta} \right\},$$

$$M_{T12} = \max_{1 \leq i < j < k \leq T} \max \left\{ \int |\theta_{ik}\theta_{jk}|^{1+\delta} dP(\xi_i) dP(\xi_j, \xi_k) \right\},$$

$$M_{T21} = \max_{1 \leq i < j < k \leq T} \max \left\{ E|\theta_{ik}\theta_{jk}|^{2(1+\delta)} \right\},$$

$$M_{T22} = \max_{1 \leq i < j < k \leq T} \max \left\{ \int |\theta_{ik}\theta_{jk}|^{2(1+\delta)} dP(\xi_i) dP(\xi_j, \xi_k) \right\},$$

$$M_{T23} = \max_{1 \leq i < j < k \leq T} \max \left\{ \int |\theta_{ik}\theta_{jk}|^{2(1+\delta)} dP(\xi_i, \xi_j) dP(\xi_k) \right\},$$

$$M_{T24} = \max_{1 \leq i < j < k \leq T} \max \left\{ \int |\theta_{ik}\theta_{jk}|^{2(1+\delta)} dP(\xi_i) dP(\xi_j) dP(\xi_k) \right\},$$

$$M_{T3} = \max_{1 \leq i < j < k \leq T} E|\theta_{ik}\theta_{jk}|^2,$$

$$M_{T4} = \max_{\substack{1 < i, j, k \leq 2T \\ i, j, k \text{ different}}} \left\{ \max_P \int |\theta_{1i}\theta_{jk}|^{2(1+\delta)} dP \right\},$$

where the maximization over P in the equation for M_{T4} is taken over the probability measures $P(\xi_1, \xi_i, \xi_j, \xi_k)$,

$P(\xi_1)P(\xi_i, \xi_j, \xi_k)$, $P(\xi_1)P(\xi_{i_1})P(\xi_{i_2}, \xi_{i_3})$, and $P(\xi_1)P(\xi_i)P(\xi_j)P(\xi_k)$,

where (i_1, i_2, i_3) is the permutation of (i, j, k) in ascending order;

$$M_{T51} = \max_{1 \leq i < j < k \leq T} \max \left\{ E \left| \int \theta_{ik}\theta_{jk}\theta_{ik}\theta_{jk} dP(\xi_i) \right|^{2(1+\delta)} \right\},$$

$$M_{T52} = \max_{1 \leq i < j < k \leq T} \max \left\{ \int \left| \int \theta_{ik}\theta_{jk}\theta_{ik}\theta_{jk} dP(\xi_i) \right|^{2(1+\delta)} dP(\xi_j) dP(\xi_k) \right\},$$

$$M_{T6} = \max_{1 \leq i < j < k \leq T} E \left| \int \theta_{ik}\theta_{jk} dP(\xi_i) \right|^2, \quad M_{T7} = \max_{1 \leq i < j < T} E \left[|\theta_{ij}|^{2(1+\delta)} \right].$$

Let

$$M_{T1} \; = \; \max_{1 \leq i \leq 2} \{M_{T1i}\} , \; M_{T2} = \max_{1 \leq i \leq 4} \{M_{T2i}\} , \; M_{T5} = \max_{1 \leq i \leq 2} \{M_{T5i}\} .$$

Assume that all the M_{Ti} are finite. Let

$$M_T \; = \; \max \left\{ T^2 M_{T1}^{\frac{1}{1+\delta}} , T^2 M_{T5}^{\frac{1}{2(1+\delta)}} , T^2 M_{T6}^{\frac{1}{2}} , T^2 M_{T7}^{\frac{1}{(1+\delta)}} \right\} ,$$

$$N_T \; = \; \max \left\{ T^{\frac{3}{2}} M_{T2}^{\frac{1}{2(1+\delta)}} , T^{\frac{3}{2}} M_{T3}^{\frac{1}{2}} , T^{\frac{3}{2}} M_{T4}^{\frac{1}{2(1+\delta)}} \right\} .$$

If $\lim_{T \to \infty} \frac{\max\{M_T, N_T\}}{\sigma_T^2} = 0$, *then*

$$\frac{1}{\sigma_T} \sum_{1 \leq s < t \leq T} \theta(\xi_s, \xi_t) \to_D N(0,1) \;\; \text{as} \;\; T \to \infty.$$

PROOF: For a given constant $0 < \rho_0 \leq \frac{1}{4}$, choose $q_T = [T^{\rho_0}] > 2$ as the largest integer part of T^{ρ_0}. Obviously, $\sum_{T=1}^{\infty} e^{-d_0 q_T} < \infty$ for any given $d_0 > 0$. Recall the notation of θ_{st} and define

$$\phi_{st} = \theta_{st} - E\left[\theta_{st} | I_{t-q}\right] \quad \text{and} \quad \psi_{st} = E\left[\theta_{st} | I_{t-q}\right]. \tag{A.2}$$

Observe that

$$L_T \; = \; \sum_{t=2}^{T} \sum_{s=1}^{t-1} \theta_{st} = \sum_{t=q+1}^{T} \sum_{s=1}^{t-q} \phi_{st} + \sum_{t=q+1}^{T} \sum_{s=1}^{t-q} \psi_{st}$$

$$+ \; \sum_{t=2}^{T} \sum_{s=t+1-q}^{t-1} \phi_{st} + \sum_{t=2}^{T} \sum_{s=t+1-q}^{t-1} \psi_{st} \equiv \sum_{j=1}^{4} L_{jT}. \tag{A.3}$$

To establish the asymptotic distribution of L_T, it suffices to show that as $T \to \infty$

$$\frac{L_{1T}}{\sigma_T} \to N(0,1) \quad \text{and} \quad \frac{L_{jT}}{\sigma_T} \to_p 0 \;\; \text{for} \;\; j = 2, 3, 4. \tag{A.4}$$

Let $V_t = \sum_{s=1}^{t-q} \phi_{st}$. Then $E[V_t | I_{t-q}] = 0$. This implies that $\{V_t\}$ is a sequence of martingale differences with respect to I_{t-q}. We now start proving the first part of (A.4). Applying a central limit theorem for martingale sequences (see Theorem 1 of Chapter VIII of Pollard 1984), in order to prove the first part of (A.4), it suffices to show that

$$\frac{1}{\sigma_T^2} \sum_{t=q+1}^{T} V_t^2 \to_p 1 \quad \text{and} \quad \frac{1}{\sigma_T^4} \sum_{t=q+1}^{T} E\left[V_t^4\right] \to 0. \tag{A.5}$$

To verify (A.5), we first need to calculate some useful quantities. Recall the definition of V_t and observe that

$$V_t^2 = \sum_{s=1}^{t-q} \phi_{st}^2 + 2 \sum_{s_1=2}^{t-q} \sum_{s_2=1}^{s_1-1} \phi_{s_1 t} \phi_{s_2 t}$$

$$\sum_{t=q+1}^{T} E[V_t^2] = \sum_{t=q+1}^{T} \sum_{s=1}^{t-q} E[\phi_{st}^2] + 2 \sum_{t=q+2}^{T} \sum_{s_1=2}^{t-q} \sum_{s_2=1}^{s_1-1} E\left[\phi_{s_1 t} \phi_{s_2 t}\right]$$

$$\equiv \sigma_{1T}^2 + \Delta_{1T}. \tag{A.6}$$

We now show that as $T \to \infty$

$$\sigma_{1T}^2 = \sigma_T^2 \left(1 + o(1)\right) \quad \text{and} \quad \Delta_{1T} = o\left(\sigma_T^2\right). \tag{A.7}$$

By Lemma A.1 (with $\eta_1 = \phi_{s_1 t}$, $\eta_2 = \phi_{s_1 t}$, $l = 2$, $p_i = 2(1 + \delta)$ and $Q = \frac{1}{1+\delta}$),

$$E\left|\phi_{s_1 t} \phi_{s_2 t}\right| \leq 10 M_{T1}^{\frac{1}{1+\delta}} \beta^{\frac{\delta}{1+\delta}}(s_1 - s_2).$$

Therefore,

$$\Delta_{1T} \leq 10 T^2 M_{T1}^{\frac{1}{1+\delta}} \sum_{i=1}^{T} \alpha^{\frac{\delta}{1+\delta}}(i) \leq C T^2 M_{T1}^{\frac{1}{1+\delta}} \tag{A.8}$$

using $\sum_{i=1}^{\infty} \alpha^{\frac{\delta}{1+\delta}}(i) < \infty$. This, together with the conditions of Theorem A.1, implies that $\Delta_{1T} = o\left(\sigma_T^2\right)$ as $T \to \infty$.

We now start to verify the first part of (A.7). Let $\sigma_{st}^2 = E[\phi_{st}^2]$. Observe that

$$E\left(\sum_{t=q+1}^{T} V_t^2 - \sigma_{1T}^2\right)^2 \leq 2E\left\{\sum_{t=q+1}^{T} \sum_{s=1}^{t-q} [\phi_{st}^2 - \sigma_{st}^2]\right\}^2$$

$$+ 8E\left\{\sum_{t=q+2}^{T} \sum_{s_1=2}^{t-q} \sum_{s_2=1}^{s_1-1} \phi_{s_1 t} \phi_{s_2 t}\right\}^2$$

$$\equiv Q_{1T} + Q_{2T}. \tag{A.9}$$

In the following, we first show that as $T \to \infty$

$$Q_{2T} = o\left(\sigma_T^4\right). \tag{A.10}$$

Using Lemma A.1 again, we can show that as $T \to \infty$

$$Q_{2T} = 8E\left\{\sum_{t=q+2}^{T} \sum_{s_1=2}^{t-q} \sum_{s_2=1}^{s_1-1} \phi_{s_1 t} \phi_{s_2 t}\right\}^2$$

$$\leq 8 \left(\sum_{t_1 \neq t_2} \sum_{s_1 \neq s_2} \sum_{r_1 \neq r_2} |E\left[\phi_{s_1 t_1} \phi_{s_2 t_1} \phi_{r_1 t_2} \phi_{r_2 t_2}\right]| \right)$$

$$\leq 8 \max\left\{M_T^2, N_T^2\right\} = o\left(\sigma_T^4\right)$$

under the conditions of Theorem A.1.

Let $C_\phi = \int \phi_{12}^2 \phi_{34}^2 dP_1(\xi_1) dP_1(\xi_2) dP_1(\xi_3) dP_1(\xi_4)$, where $P_1(\xi_i)$ denotes the probability measure of ξ_i.

Using Lemma A.1 repeatedly, we have that for different i, j, k, l

$$\left|E\left[\phi_{ij}^2 \phi_{kl}^2\right] - C_\phi\right| \leq 10\left\{\alpha(\Delta(i,j,k,l))\right\}^{1 - \frac{1}{1+\delta}} M_{T4}^{\frac{1}{1+\delta}}$$

$$= 10 M_{T4}^{\frac{1}{1+\delta}} \left\{\alpha(\Delta(i,j,k,l))\right\}^{\frac{\delta}{1+\delta}}, \quad (A.11)$$

where $\Delta(i,j,k,l)$ is the minimum increment in the sequence which is the permutation of i, j, k, l in ascending order.

Similarly to (A.11), we can have for all different i, j, k, l

$$\left|\sigma_{ij}^2 \sigma_{kl}^2 - C_\phi\right| \leq 10 M_{T4}^{\frac{1}{1+\delta}} \left\{\alpha(\Delta(i,j,k,l))\right\}^{\frac{\delta}{1+\delta}}. \quad (A.12)$$

Therefore, using (A.11) and (A.12),

$$Q_{1T} = 2E\left\{ \sum_{t=q+2}^{T} \sum_{s=1}^{t-q} \left[\phi_{st}^2 - \sigma_{st}^2\right] \right\}^2$$

$$\leq 2 \left(\sum_{t_1, t_2} \sum_{s_1, s_2} \left|E\left[\phi_{ij}^2 \phi_{kl}^2\right] - \sigma_{ij}^2 \sigma_{kl}^2\right| \right)$$

$$\leq 2 \left(\sum_{t_1, t_2} \sum_{s_1, s_2} \left|E\left[\phi_{ij}^2 \phi_{kl}^2\right] - C_\phi\right| + \left|C_\phi - \sigma_{ij}^2 \sigma_{kl}^2\right| \right)$$

$$\leq \left\{ O\left(T^3 M_{T4}^{\frac{1}{1+\delta}}\right) + O\left(T^3 M_{T3}\right) \right\} = o(\sigma_T^4). \quad (A.13)$$

It now follows from (A.9)–(A.13) that for any $\epsilon > 0$

$$P\left\{ \left| \frac{1}{\sigma_{1T}^2} \sum_{t=q+1}^{T} V_t^2 - 1 \right| \geq \epsilon \right\} \leq \frac{1}{\sigma_{1T}^4 \epsilon^2} E\left[\sum_{t=q+1}^{T} V_t^2 - \sigma_{1T}^2 \right]^2 \to 0. \quad (A.14)$$

Thus, the first part of (A.5) is proved. Note that for $q + 1 \leq k \leq T$,

$$E[V_k^4] = E\left\{ \sum_{i=1}^{k-q} \phi_{ik}^2 + 2 \sum_{1 \leq i < j < k-q} \phi_{ik} \phi_{jk} \right\}^2$$

$$
= \quad E \left\{ \sum_{i=1}^{k-q} \phi_{ik}^4 + 6 \sum_{1 \le i < j < k-q} \phi_{ik}^2 \phi_{jk}^2 + 4 \sum_{l=1}^{k-q} \sum_{1 \le i < j < k-q} \phi_{lk}^2 \phi_{ik} \phi_{jk} \right\}
$$

$$
+ \quad 4E \left\{ \sum_{q \le i < j < k, 1 \le s < t < k-q, (i,j) \ne (s,t)} \phi_{ik} \phi_{jk} \phi_{sk} \phi_{tk} \right\}
$$

$$
= \quad 4 \sum_{l=1}^{k-q} \sum_{1 \le i < j < k-q} E \left[\phi_{lk}^2 \phi_{ik} \phi_{jk} \right]
$$

$$
+ \quad 4 \sum_{1 \le i < j < k-q, \ 1 \le s < t < k-q, (i,j) \ne (s,t)} E \left[\phi_{ik} \phi_{jk} \phi_{sk} \phi_{tk} \right]
$$

$$
+ \quad O \left(T^2 M_{T3} \right). \tag{A.15}
$$

It is easy to see that

$$
\int |\phi_{ik} \phi_{jk} \phi_{sk} \phi_{tk}|^{1+\delta} \, dP \quad \le \quad \sqrt{\int |\phi_{ik} \phi_{jk}|^{2(1+\delta)} \, dP \int |\phi_{sk} \phi_{tk}|^{2(1+\delta)} \, dP}
$$

$$
\le \quad M_{T4}.
$$

Similarly to (A.11), we can have for any $(i,j) \ne (s,t)$,

$$
|E[\phi_{ik} \phi_{jk} \phi_{sk} \phi_{tk}]| \le 10 M_{T4}^{\frac{1}{1+\delta}} \left\{ \alpha(\Delta(i,j,s,t)) \right\}^{\frac{\delta}{1+\delta}}, \tag{A.16}
$$

where $\Delta(\cdot)$ is as defined before.

Consequently,

$$
\sum_{k=q+1}^{T} E[V_k^4] = O \left(T^3 M_{T4}^{\frac{1}{1+\delta}} \right) = o(\sigma_T^4). \tag{A.17}
$$

This finishes the proof of the first part of (A.5) and therefore the proof of (A.5).

Applying Lemmas A.1 and A.2 implies that as $T \to \infty$

$$
E |L_{2T}| \quad \le \quad \sum_{t=q+1}^{T} \sum_{s=1}^{t-q} E \left| E \left[\theta_{st} | I_{t-q} \right] \right|
$$

$$
\le \quad C \sum_{t=q+1}^{T} \sum_{s=1}^{t-q} \alpha^{\frac{\delta}{1+\delta}} (t - q - s) \, M_{T7}^{\frac{1}{2(1+\delta)}} \le C \left(T M_{T7}^{\frac{1}{2(1+\delta)}} \right)
$$

$$
= \quad o(\sigma_T) \tag{A.18}
$$

using the conditions of Theorem A.1.

The second part of (A.4) for L_{4T} follows from the conditions of Theorem A.1 and

$$
\begin{aligned}
E\left|L_{4T}\right| &\leq \sum_{t=2}^{T} \sum_{s=t+1-q}^{t-1} E\left(E\left[\left|\theta_{st}\right| \left|I_{t-q}\right]\right)\right. \\
&\leq C \sum_{t=2}^{T} \sum_{s=t+1-q}^{t-1} \alpha^{\frac{\delta}{1+\delta}}(t-1-s) M_{T7}^{\frac{1}{2(1+\delta)}} \leq C\left(T M_{T7}^{\frac{1}{2(1+\delta)}}\right) \\
&= o(\sigma_T).
\end{aligned}
\tag{A.19}
$$

We finally prove the second part of (A.4) for L_{3T}. Similarly, using Lemma A.1, we can show that as $T \to \infty$

$$
\left|\sum_{t=2}^{T} \sum_{s_1=t+1-q}^{t-1} \sum_{s_2 \neq s_1, s_2=t+1-q}^{t-1} E\left[\phi_{s_1 t} \phi_{s_2 t}\right]\right| \leq \sum_{t=2}^{T} \sum_{s_1=t+1-q}^{t-1} \sum_{s_2 \neq s_1, s_2=t+1-q}^{t-1}
$$
$$
\times E\left[\left|\phi_{s_1 t} \phi_{s_2 t}\right|\right]
$$
$$
\leq o\left(T^2 q M_{T3}\right),
\tag{A.20}
$$

$$
\left|\sum_{t_1=3}^{T} \sum_{t_2=t_1+1-q}^{t_1-1} \sum_{s_1=t_1+1-q}^{t_1-1} \sum_{s_2=t_2+1-q}^{t_2-1} E\left[\phi_{s_1 t_1} \phi_{s_2 t_2}\right]\right| \leq o\left(T^2 q^2 M_{T3}\right).
\tag{A.21}
$$

Using (A.20) and (A.21) implies that as $T \to \infty$

$$
\begin{aligned}
E\left[L_{3T}^2\right] &= \sum_{t=2}^{T} \sum_{s=t+1-q}^{t-1} E\left[\phi_{st}^2\right] + \sum_{t=2}^{T} \sum_{s_1=t+1-q}^{t-1} \sum_{s_2 \neq s_1, s_2=t+1-q}^{t-1} E\left[\phi_{s_1 t} \phi_{s_2 t}\right] \\
&\quad + 2 \sum_{t_1=q+2}^{T} \sum_{t_2=2}^{t_1-q} \sum_{s_1=t_1+1-q}^{t_1-1} \sum_{s_2=t_2+1-q}^{t_2-1} E\left[\phi_{s_1 t_1} \phi_{s_2 t_2}\right] \\
&\quad + 2 \sum_{t_1=2}^{T} \sum_{t_2=t_1+1-q}^{t_1-1} \sum_{s_1=t_1+1-q}^{t_1-1} \sum_{s_2=t_2+1-q}^{t_2-1} E\left[\phi_{s_1 t_1} \phi_{s_2 t_2}\right] \\
&= \sum_{t=2}^{T} \sum_{s=t+1-q}^{t-1} E\left[\phi_{st}^2\right] + \sum_{t=2}^{T} \sum_{s_1=t+1-q}^{t-1} \sum_{s_2 \neq s_1, s_2=t+1-q}^{t-1} E\left[\phi_{s_1 t} \phi_{s_2 t}\right] \\
&\quad + 2 \sum_{t_1=3}^{T} \sum_{t_2=t_1+1-q}^{t_1-1} \sum_{s_1=t_1+1-q}^{t_1-1} \sum_{s_2=t_2+1-q}^{t_2-1} E\left[\phi_{s_1 t_1} \phi_{s_2 t_2}\right] \\
&= O\left(T^{\frac{3}{2}} M_{T3}^{\frac{1}{2}}\right) + O\left(T q^2 M_{T3}^{\frac{1}{2}}\right) = o\left(\sigma_T^2\right)
\end{aligned}
\tag{A.22}
$$

in view of the fact that the third term of (A.22) is zero because of

$$
E[\phi_{s_2 t_2} \phi_{s_1 t_1} | I_{t_1-q}] = 0.
$$

This completes the proof of Theorem A.1.

In order to establish some useful lemmas without including non–essential technicality, we introduce the following simplified notation:

$$a_{st} = \frac{1}{T\sqrt{h^d}\sigma_0} K\left(\frac{X_s - X_t}{h}\right), \ Q_T(h) = \sum_{t=1}^{T}\sum_{s=1,\neq t}^{T} a_{st}e_s e_t,$$

$$\rho(h) = \frac{\sqrt{2}K^{(3)}(0)\int \pi^3(u)du}{3}\left(\sqrt{\int \pi^2(u)du \int K^2(v)dv}\right)^{-3}\sqrt{h^d},$$

where $\sigma_0^2 = 2\mu_2^2\,\nu_2\int K^2(v)dv$ with $\nu_2 = E[\pi^2(X_1)]$ and $\mu_2 = E[e_1^2]$, and $K^{(3)}(\cdot)$ denotes the three–time convolution of $K(\cdot)$ with itself. We now have the following theorem.

Theorem A.2. *Suppose that the conditions of Theorem 3.1 hold. Then for any h*

$$\sup_{x\in R^1}\left|P\left(Q_T(h) \leq x\right) - \Phi(x) + \rho(h)\left(x^2 - 1\right)\phi(x)\right| = O\left(h^d\right). \quad (A.23)$$

PROOF: The proof is based on a nontrivial application of Lemma A.3. As the proof itself is extremely technical, we provide only an outline below.

In view of $Q_T(h)$, we need to follow the proofs of Theorems 1.1 and 3.1 as well as Lemmas 3.2–3.5 of Götze, Tikhomirov and Yurchenko (2004) step-by-step to finish the proof of Theorem A.2. Note that the proofs of their Theorems 1.1 and 3.1 remain true. The proofs of their Lemmas 3.2–3.5 also remain true by successive conditioning arguments when needed.

Alternatively, we may apply Lemma A.3 to the conditional probability $P\left(Q_T(h) \leq x|\mathcal{X}_T\right)$ and then use the dominated convergence theorem to deduce (A.23) unconditionally. To avoid repeating the conditioning argument (given $\mathcal{X}_T$) for each case in the following derivations, the corresponding conditioning arguments are all understood to be held in probability with respect to the joint distribution of $\mathcal{X}_T = (X_1, \cdots, X_T)$.

In any case, in order to apply Lemma A.3, we need to verify certain conditions of Lemma A.3.

$$a_{jj} = T^{-1}h^{-d/2}K(0), \ \mathbf{d}^\tau = T^{-1}h^{-d/2}K(0)\,(1,\cdots,1)^\tau$$

$$\mathbf{Tr}\mathbf{A} = h^{-d/2}K(0), \ V^2 = (Th^d)^{-1}K^2(0)$$

$$\|\mathbf{A_0}\|^2 = T^{-2}h^{-d}\sum_{\substack{s,t=1\\s\neq t}}^{T} K^2\left(\frac{X_s - X_t}{h}\right)$$

$$\mathbf{d}^\tau \mathbf{A_0}\mathbf{d} = T^{-3}h^{-3d/2}K^2(0)\sum_{\substack{s,t=1\\s\neq t}}^{T} K\left(\frac{x_s - x_t}{h}\right). \quad (A.24)$$

Obviously,

$$\mathbf{Tr}(\mathbf{A}_0^3) = \sum_{\substack{q=1 \\ k \neq q \\ j \neq q}}^{T} \sum_{\substack{k=1 \\ j \neq k}}^{T} \sum_{j=1}^{T} a_{qk} a_{kj} a_{jq} = \left(T^{-1} h^{-d/2}\right)^3 \sum_{\substack{q=1 \\ k \neq q \\ j \neq q}}^{T} \sum_{\substack{k=1 \\ j \neq k}}^{T} \sum_{\substack{j=1 \\ j \neq q}}^{T}$$

$$\times \ K\left(\frac{X_q - X_k}{h}\right) K\left(\frac{X_k - X_j}{h}\right) K\left(\frac{X_j - X_q}{h}\right).$$

Using the stationary ergodic theorem, the sums involving the kernel function K in (A.24) can be approximated as follows:

$$\frac{1}{T^2} \sum_{\substack{j,k=1 \\ j \neq k}}^{T} K^2\left(\frac{X_i - X_j}{h}\right) \ \approx \ \int \int K^2\left(\frac{x-y}{h}\right) \pi(x,y) dx dy$$

$$\approx \ h^d \int \int K^2(u) \pi(y+uh, y) du dy$$

$$\approx \ h^d \int \int K^2(u) \pi^2(y) du dy$$

$$\approx \ h^d \int \int K^2(u) \pi^2(v) du dv, \qquad \text{(A.25)}$$

where $\pi(x,y)$ denotes the joint density function of $(X_1, X_{1+\tau})$ for any $\tau \geq 1$, and $\pi(x)$ is the marginal density function of X_1.

Similarly, for the second sum in expression (A.24)

$$\frac{1}{T^2} \sum_{\substack{s,t=1 \\ s \neq t}}^{T} K\left(\frac{X_s - X_t}{h}\right) \approx h^d \int K(u) du \int \pi^2(v) dv. \qquad \text{(A.26)}$$

For the triple sum in expression (A.25) we find

$$\frac{1}{T^3} \sum_{\substack{q=1 \\ k \neq q \\ j \neq q}}^{T} \sum_{\substack{k=1 \\ j \neq k}}^{T} \sum_{\substack{j=1 \\ j \neq q}}^{T} K\left(\frac{X_q - X_k}{h}\right) K\left(\frac{X_k - X_j}{h}\right) K\left(\frac{X_j - X_q}{h}\right)$$

$$\approx \int \int \int K\left(\frac{x-y}{h}\right) K\left(\frac{y-z}{h}\right) K\left(\frac{z-x}{h}\right) \pi(x,y,z) dx dy dz$$

$$\approx h^{2d} \int \int \int K(-(u+v)) K(v) K(u) \pi(z-uh, z+vh, z) du dv dz$$

$$\approx h^{2d} \int \int \int K(-(u+v)) K(v) K(u) \pi^3(z) du dv dz$$

$$= h^{2d} \int \int K(u+v) K(v) K(u) du dv$$

$$= h^{2d} \int \left(\int K(w)K(w - v)dw \right) K(v)dv$$

$$= h^{2d} \int K * K(v)K(v)dv \int \pi^3(u)du$$

$$= h^{2d} (K * K * K)(0) \int \pi^3(u)du, \qquad (A.27)$$

where $\pi(x, y, z)$ denotes the joint density of $(X_1, X_{1+\tau_1}, X_{1+\tau_2})$ for any $\tau_1, \tau_2 \geq 1$.

Combining (A.24)—(A.27) we obtain the following behaviours

$$\mathbf{Tr A} \approx h^{-d/2}K(0), \quad V^2 \approx T^{-1}h^{-d}K^2(0)$$

$$\|\mathbf{A_0}\|^2 \approx \int K^2(u)du \int \pi^2(v)dv$$

$$\mathbf{d}^\tau \mathbf{A_0 d} \approx T^{-1}h^{-d/2}K^2(0) \int K(u)du \int \pi^2(v)dv$$

$$\mathbf{Tr}(\mathbf{A_0^3}) \approx h^{d/2} K^{(3)}(0) \int \pi^3(u)du, \qquad (A.28)$$

where $K^{(3)}(\cdot) = (K * K * K)(\cdot)$ is the three times convolution of K with itself.

From this we get approximations for the quantities σ_*^2 and κ involved in Lemma A.3:

$$\sigma_*^2 \approx T^{-1}h^{-d}(\mu_4 - \mu_2^2)K^2(0) + 2\mu_2^2 \int K^2(u)du \int \pi^2(v)dv$$

$$\kappa \approx \frac{\frac{\mu_3^2 K^2(0)}{Th^d} + \frac{4\mu_2^3 \int \pi^3(u)du}{3}K^{(3)}(0)}{\sigma_*^3} \sqrt{h^d}$$

$$\approx \frac{\sqrt{2}K^{(3)}(0)}{3} \left(\sqrt{\int K^2(u)du} \right)^{-3} c(\pi) \sqrt{h^d} \equiv \rho(h),$$

where $c(\pi) = \dfrac{\int \pi^3(x)dx}{\left(\sqrt{\int \pi^2(x)dx} \right)^3}$

In order to apply Lemma A.3 to finish the proof, we need to show that the upperbound of Lemma A.3 tends to 0 as $T \to \infty$. Observe that

$$\|\mathbf{A}\|^2 = \|\mathbf{A_0}\|^2 + \sum_{j=1}^{T} a_{jj}^2$$

$$\approx \int K^2(u)du \int \pi^2(v)dv + (nh^d)^{-1}K^2(0). \qquad (A.29)$$

Similarly to (A.27), we may show that

$$
\sum_{t=1}^{T} \mathcal{L}_t^4 = \sum_{s=1}^{T} \left(\sum_{t=1}^{T} a_{st}^2 \right)^2 = \sum_{s=1}^{T} \sum_{t=1}^{T} a_{st}^4
$$

$$
+ \sum_{s=1}^{T} \sum_{t_1=1}^{T} \sum_{t_2=1,\neq t_1}^{T} a_{st_1}^2 a_{st_2}^2 = \frac{1}{T^2 h^d} \int K^4(u)du \int \pi^2(v)dv
$$

$$
+ \frac{1}{Th^d} \int \int K^2(w)K^2\left(w + \frac{u-v}{h}\right) \pi(u,v)dwdudv
$$

$$
= \frac{1}{T^2 h^d} \int K^4(u)du \int \pi^2(v)dv
$$

$$
+ \frac{1}{T} K_2^{(2)}(0) \int \pi^2(v)dv, \tag{A.30}
$$

where $K_2^{(2)}(0)$ is the two–time convolution of $K^2(\cdot)$ with itself.

Similarly, we may show that as $T \to \infty$

$$
\lambda_1 \leq \max_{1 \leq j \leq T} \sum_{i=1}^{T} |a_{ij}| \leq \sqrt{h^d} \int K(u)du \int \pi^2(v)dv. \tag{A.31}
$$

Consequently, using that $h \to 0$ and $Th^d \to \infty$, we find that

$$
\frac{|\lambda_1|^2}{\|\mathbf{A}\|^2} \approx \frac{h^d \left(\int \pi^2(v)dv \right)^2}{\int K^2(u)du \int \pi^2(v)dv}. \tag{A.32}
$$

From (A.27)–(A.30) we then find that

$$
\frac{\left(\sum_{t=1}^{T} \mathcal{L}_t^4 \right)^{1/2}}{\|\mathbf{A}\|^2} \approx \frac{\sqrt{K_2^{(2)}(0) \int \pi^2(v)dv}}{\sqrt{T}}. \tag{A.33}
$$

Thus, (A.32) and (A.33) imply that there is some constant C_∞ such that

$$
M \approx C_\infty h^d, \tag{A.34}
$$

which shows that the upperbound in Lemma A.3 tends to 0 at a rate proportional to h^d. This completes the proof of Theorem A.2.

References

Agarwal, G. G., and Studden, W. J., 1980. Asymptotic integrated mean square error using least squares and bias minimizing splines. *Annals of Statistics* 8, 1307–1325.

Ahn, D. H., and Gao, B., 1999. A parametric nonlinear model of term structure dynamics. *Review of Financial Studies* 12, 721–762.

Aït-Sahalia, Y., 1996a. Nonparametric pricing of interest rate derivative securities. *Econometrica* 64, 527–560.

Aït-Sahalia, Y., 1996b. Testing continuous-time models of the spot interest rate. *Review of Financial Studies* 9, 385–426.

Aït-Sahalia, Y., 1999. Transition densities for interest rate and other nonlinear diffusions. *Journal of Finance* 54, 1361–1395.

Aït-Sahalia, Y., Bickel, P., and Stoker, T., 2001. Goodness-of-fit tests for regression using kernel methods. *Journal of Econometrics* 105, 363–412.

Aït-Sahalia, Y., and Lo, A., 1998. Nonparametric estimation of state–price densities implicit in financial asset prices. *Journal of Finance* 53, 499–547.

Aït-Sahalia, Y., and Lo, A., 2000. Nonparametric risk management and implied risk aversion. *Journal of Econometrics* 94, 9–51.

Akaike, H., 1973. Information theory and an extension of the maximum likelihood principle. In *2nd International Symposium on Information Theory* (Edited by B. N. Petrov and F. Csáki), 267–281. Akadémiai Kiado, Budapest.

Andersen, T. G., and Lund, J., 1997. Estimation in continuous–time stochastic volatility models of the short term interest rate. *Journal of Econometrics* 77, 343–378.

Andersen, T. G., and Sørensen, B. E., 1996. GMM estimation of a stochastic volatility model: a Monte Carlo study. *Journal of Business and Economic Statistics* 14, 328–352.

Andrews, D. W., 1991. Asymptotic normality of series estimators for nonparametric and semiparametric regression models. *Econometrica* 59, 307–345.

Andrews, D. W., 1993. Tests for parameter instability and structural change with unknown change point. *Econometrica* 61, 821–856.

Andrews, D. W., 1997. A conditional kolmogorov test. *Econometrica* 65, 1097–1028.

Andrews, D. W., and Ploberger, W., 1994. Optimal tests when a nuisance parameter is present only under the alternative. *Econometrica* 62, 1383–1414.

Anh, V., Angulo, J., and Ruiz–Medina, M., 1999. Possible long–range dependence in fractional random fields. *Journal of Statistical Planning & Inference* 80, 95–110.

Anh, V., and Heyde, C. (ed.), 1999. Special issue on long–range dependence. *Journal of Statistical Planning & Inference* 80, 1.

Anh, V., and Inoue, A., 2005. Financial markets with memory I: dynamic models. *Stochastic Analysis & Its Applications* 23, 275–300.

Anh, V., Inoue, A., and Kasahara, Y., 2005. Financial markets with memory II: innovation processes and expected utility maximization. *Stochastic Analysis & Its Applications* 23, 301–328.

Anh, V., Wolff, R. C. L., Gao, J., and Tieng, Q., 1999. Local linear regression with long–range dependent errors. *Australian & New Zealand Journal of Statistics* 41, 463–479.

Arapis, M., and Gao, J., 2006. Empirical comparisons in short–term interest rate models using nonparametric methods. *Journal of Financial Econometrics* 4, 310–345.

Auestad, B., and Tjøstheim, D., 1990. Identification of nonlinear time series: first order characterization and order determination. *Biometrika* 77, 669–687.

Avramidis, P., 2005. Two–step cross–validation selection method for partially linear models. *Statistica Sinica* 15, 1033–1048.

Bachelier, L., 1900. Théorie e la Speculation. Reprinted in Cootner (ed.), 17–78.

Baillie, R., and King, M. L. (eds.), 1996. Special Issue of the *Journal of Econometrics. Annals of Econometrics* 73.

Bandi, F., and Phillips, P. C. B., 2003. Fully nonparametric estimation of scalar diffusion models. *Econometrica* 71, 241–283.

Beran, J., 1994. *Statistics for Long Memory Processes*. Chapman and Hall, London.

Beran, J., and Feng, Y., 2002. Local polynomial fitting with long-memory, short-memory and antipersistent errors. *Annals of the Institute of Statistical Mathematics* 54, 291–311.

Beran, J., and Ghosh, S., 1998. Root–n–consistent estimation in partial linear models with long–memory errors. *Scandinavian Journal of Statistics* 25, 345–357.

Beran, J., Ghosh, S., and Sibbertsen, P., 2003. Nonparametric M-estimation with long–memory errors. *Journal of Statistical Planning & Inference* 117, 199–205.

Berkes, I., Horváth, L., Kokoszka, P., and Shao, Q., 2006. On discriminating between long–range dependence and changes in mean. *Annals of Statistics* 34, 1116–1140.

Bickel, P., and Zhang, P., 1992. Variable selection in nonparametric regression with categorical covariates. *Journal of the American Statistical Association* 87, 90–97.

Black, F., and Scholes, M., 1973. The pricing of options and corporate liabilities. *Journal Political Economy* 3, 637–654.

Bloomfield, P., 1976. *Fourier Analysis of Time Series: An Introduction.* John Wiley, New York.

Boente, G., and Fraiman, R., 1988. Consistency of a nonparametric estimate of a density function for dependent variables. *Journal of Multivariate Analysis* 25, 90–99.

Breidt, F. J., Crato, N. and de Lima, P. J. F., 1998. The detection and estimation of long–memory in stochastic volatility. *Journal of Econometrics* 83, 325–348.

Brennan, M., and Schwartz, E., 1980. A continuous–time approach to the pricing of bonds. *Journal of Banking and Finance* 3, 133–145.

Broto, C., and Ruiz, E., 2004. Estimation methods for stochastic volatility models: a survey. *Journal of Economic Surveys* 18, 613–649.

Brockwell, P., and Davis, R., 1990. *Time Series: Theory and Methods.* Springer, New York.

Cai, Z., Fan, J., and Li, R., 2000. Efficient estimation and inferences for varying–coefficient models. *Journal of the American Statistical Association* 95, 888–902.

Cai, Z., Fan, J., and Yao, Q., 2000. Functional-coefficient regression models for nonlinear time series. *Journal of the American Statistical Association* 95, 941–956.

Cai, Z., and Hong, Y., 2003. Nonparametric methods in continuous–time finance: a selective review. In *Recent Advances and Trends in*

Nonparametric Statistics (Edited by M. G. Akritas and D. Politis), 283–302. North–Holland, Amsterdam.

Carroll, R. J., Fan, J., Gijbels, I., and Wand, M. P., 1997. Generalized partially linear single–index models. *Journal of the American Statistical Association* 92, 477–489.

Casas, I., and Gao, J., 2005. Specification testing in semiparametric continuous–time diffusion models: theory and practice. Working paper available from www.maths.uwa.edu.au/~jiti/casgao05.pdf.

Casas, I., and Gao, J., 2006. Estimation in continuous–time stochastic volatility models with long–range dependence. Working paper available from www.maths.uwa.edu.au/~jiti/cgjbes.pdf.

Chan, K., Karolyi, F., Longstaff, F., and Sanders, A., 1992. An empirical comparison of alternative models of the short–term interest rate. *Journal of Finance* 47, 1209–1227.

Chan, N. H., 2002. *Time Series: Applications to Finance.* Wiley Interscience, Hoboken, New Jersey.

Chapman, D., and Pearson, N., 2000. Is the short rate drift actually nonlinear ? *Journal of Finance* 54, 355–388.

Chen, H., and Chen, K., 1991. Selection of the splined variables and convergence rates in a partial spline model. *Canadian Journal of Statistics* 19, 323–339.

Chen, R., Liu, J., and Tsay, R., 1995. Additivity tests for nonlinear autoregression. *Biometrika* 82, 369–383.

Chen, R., and Tsay, R., 1993. Nonlinear additive ARX models. *Journal of the American Statistical Association* 88, 955–967.

Chen, S. X., and Gao, J., 2004. An adaptive empirical likelihood test for parametric time series regression. *Journal of Econometrics* (forthcoming and available from www.maths.uwa.edu.au/~jiti/cg04.pdf).

Chen, S. X., and Gao, J., 2005. On the use of the kernel method for specification tests of diffusion models. Working paper available from www.maths.uwa.edu.au/~jiti/cg05.pdf.

Chen, S. X., Härdle, W., and Li, M., 2003. An empirical likelihood goodness–of–fit test for time series. *Journal of the Royal Statistical Society Series B* 65, 663–678.

Chen, X., and Fan, Y., 1999. Consistent hypothesis testing in semiparametric and nonparametric models for econometric time series. *Journal of Econometrics* 91, 373–401.

Chen, X., and Fan, Y., 2005. Pseudo–likelihood ratio tests for model selection in semiparametric multivariate copula models. *Canadian Journal of Statistics* 33, 389–414.

Cheng, B., and Robinson, P. M., 1994. Semiparametric estimation from time series with long–range dependence. *Journal of Econometrics* 64, 335–353.

Cheng, B., and Tong, H., 1992. On consistent nonparametric order determination and chaos. *Journal of the Royal Statistical Society Series B* 54, 427–449.

Cheng, B., and Tong, H., 1993. Nonparametric function estimation in noisy chaos. In *Developments in Time Series Analysis* (Edited by T. Subba Rao), 183–206. Chapman and Hall, London.

Comte, F., 1996. Simulation and estimation of long memory continuous time models. *Journal of Time Series Analysis* 17, 19–36.

Comte, F., and Renault, E., 1996. Long memory in continuous-time models. *Journal of Econometrics* 73, 101–149.

Comte, F., and Renault, E., 1998. Long memory in continuous–time stochastic volatility models. *Mathematical Finance* 8, 291–323.

Corradi, V., and Swanson, N. R., 2005. Bootstrap specification tests for diffusion processes. *Journal of Econometrics* 124, 117–148.

Corradi, V., and White, H., 1999. Specification tests for the variance of a diffusion process. *Journal of Time Series Analysis* 20, 253–270.

Cox, J., Ingersoll, E., and Ross, S., 1985. An intertemporal general equilibrium model of asset prices. *Econometrica* 53, 363–384.

Dahlhaus, R., 1989. Efficient parameter estimation for self–similar processes. *Annals of Statistics* 17, 1749–1766.

Delgado, M. A., and Hidalgo, J., 2000. Nonparametric inference on structural breaks. *Journal of Econometrics* 96, 113–144.

Deo, R., and Hurvich, C. M., 2001. On the log–periodogram regression estimator of the memory parameter in long memory stochastic volatility models. *Econometric Theory* 17, 686–710.

Dette, H., 1999. A consistent test for the functional form of a regression based on a difference of variance estimators. *Annals of Statistics* 27, 1012–1040.

Dette, H., 2002. A consistent test for heteroscedasticity in nonparametric regression based on the kernel method. *Journal of Statistical Planning & Inference* 103, 311–329.

Dette, H., and Von Lieres und Wilkau, C., 2001. Testing additivity by kernel-based methods—what is a reasonable test ? *Bernoulli* 7, 669–697.

Dette, H., and Von Lieres und Wilkau, C., 2003. On a test for a parametric form of volatility in continuous time financial models. *Finance Stochastics* 7, 363–384.

DeVore, R. A., and Lorentz, G. G., 1993. *Constructive Approximation.* Springer, New York.

Dickey, D. A., and Fuller, W. A., 1979. Distribution of estimators for autoregressive time series with a unit root. *Journal of the American Statistical Association* 74, 427–431.

Ding, Z., and Granger, C. W. J., 1996. Modelling volatility persistence of speculative returns: a new approach. *Journal of Econometrics* 73, 185–215.

Ding, Z., Granger, C. W. J., and Engle, R., 1993. A long memory property of stock market returns and a new model. *Journal of Empirical Finance* 1, 83–105.

Dong, C., Gao, J., and Tong, H., 2006. Semiparametric penalty function method in partially linear model selection. Forthcoming in *Statistica Sinica* (available from www.maths.uwa.edu.au/~jiti/dgt.pdf).

Doukhan, P., 1995. *Mixing–Properties and Examples.* Lecture Notes in Statistics. Springer–Verlag, New York.

Durham, G. B., 2004. Likelihood specification analysis of continuous–time models of the short–term interest rate. Forthcoming in the *Journal of Financial Economics.*

Eastwood, B., and Gallant, R., 1991. Adaptive truncation rules for seminonparametric estimates achieving asymptotic normality. *Econometric Theory* 7, 307–340.

Engle, R. F., 2001. Financial econometrics–a new dicipline with new methods. *Journal of Econometrics* 100, 53–56.

Engle, R. F., and Granger, C. W. J., 1987. Co–integration and error correction: representation, estimation and testing. *Econometrica* 55, 251–276.

Engle, R. F., Granger, C. W. J., Rice, J. A., and Weiss, A., 1986. Semiparametric estimates of the relation between weather and electricity sales. *Journal of the American Statistical Association* 81, 310–320.

Eubank, R. L., 1988. *Spline Smoothing and Nonparametric Regression.* Marcel Dekker, Inc., New York.

Eubank, R. L., 1999. *Nonparametric Regression and Spline Smoothing.* Marcel Dekker, Inc., New York.

Eubank, R. L., and Hart, J. D., 1992. Testing goodness-of-fit in regression via order selection. *Annals of Statistics* 20, 1412–1425.

Eubank, R. L., and Spiegelman, C. H., 1990. Testing the goodness of fit of a linear model via nonparametric regression techniques. *Journal of the American Statistical Association* 85, 387–392.

Eumunds, D. E., and Moscatelli, V. B., 1977. Fourier approximation and embeddings of Sobolev space. Dissertationae Mathematicae. Polish Scientific Publishers, Warsaw.

Fan, J., 1996. Test of significance based on wavelet thresholding and Neyman's truncation. *Journal of the American Statistical Association* 91, 674–688.

Fan, J., 2005. A selective overview of nonparametric methods in financial econometrics. With comments and a rejoinder by the author. *Statistical Science* 20, 317–357.

Fan, J., and Gijbels, I., 1996. *Local Polynomial Modelling and Its Applications*. Chapman and Hall, London.

Fan, J., Härdle, W., and Mammen, E., 1998. Direct estimation of low dimensional components in additive models. *Annals of Statistics* 26, 943–971.

Fan, J., and Huang, L. S., 2001. Goodness-of-fit tests for parametric regression models. *Journal of the American Statistical Association* 453, 640–652.

Fan, J., and Huang, T., 2005. Profile likelihood inferences on semiparametric varying–coefficient partially linear models. *Bernoulli* 11, 1031–1057.

Fan, J., and Jiang, J., 2005. Nonparametric inferences for additive models. *Journal of the American Statistical Association* 100, 890–907.

Fan, J., Jiang, J., Zhang, C., and Zhou, Z., 2003. Time–dependent diffusion models for term structure dynamics. *Statistica Sinica* 13, 965–992.

Fan, J., and Li, R., 2001. Variable selection via nonconcave penalized likelihood and its oracle properties. *Journal of the American Statistical Association* 96, 1348–1360.

Fan, J., and Li, R., 2002. Variable selection for Cox's proportional hazards model and frailty model. *Annals of Statistics* 30, 74–99.

Fan, J., and Yao, Q., 1998. Efficient estimation of conditional variance functions in stochastic regression. *Biometrika* 85, 645–660.

Fan, J., and Yao, Q., 2003. *Nonlinear Time Series: Nonparametric and Parametric Methods*. Springer, New York.

Fan, J., and Zhang, C., 2003. A re-examination of Stanton's diffusion estimation with applications to financial model validation. *Journal of the American Statistical Association* 461, 118–134.

Fan, J., Zhang, C. M., and Zhang, J., 2001. Generalized likelihood ratio statistics and Wilks phenomenon. *Annals of Statistics* 29, 153–193.

Fan, Y., and Li, Q., 1996. Consistent model specification tests: omitted variables and semiparametric functional forms. *Econometrica* 64, 865–890.

Fan, Y., and Li, Q., 1999. Central limit theorem for degenerate U–statistics of absolutely regular processes with applications to model specification testing. *Journal of Nonparametric Statistics* 10, 245–271.

Fan, Y., and Li, Q., 2000. Consistent model specification tests: kernel-based tests versus Bierens' ICM tests. *Econometric Theory* 16 1016–1041.

Fan, Y., and Li, Q., 2003. A kernel-based method for estimating additive partially linear models. *Statistica Sinica* 13, 739–762.

Fan, Y., and Linton, O., 2003. Some higher–theory for a consistent nonparametric model specification test. *Journal of Statistical Planning and Inference* 109, 125–154.

Fox, R., and Taqqu, M. S., 1986. Large–sample properties of parameter estimates for strongly dependent stationary Gaussian time series. *Annals of Statistics* 14, 512–532.

Franke, J., Kreiss, J. P., and Mammen, E., 2002. Bootstrap of kernel smoothing in nonlinear time series. *Bernoulli* 8, 1–38.

Franses, P. H., and van Dijk, D., 2000. *Nonlinear Time Series Models in Empirical Finance*. Cambridge University Press.

Frisch, U., 1995. *Turbulence*. Cambridge University Press.

Galka, A., 2000. *Topics in Nonlinear Time Series Analysis with Implications for EEG Analysis*. World Scientific, Singapore.

Gallant, A. R., 1981. On the bias in flexible functional forms and an essentially unbiased form: the Fourier flexible form. *Journal of Econometrics* 15, 211–245.

Gallant, A. R., and Souza, G., 1991. On the asymptotic normality of Fourier flexible form estimates. *Journal of Econometrics* 50, 329–353.

Gao, J., 1998. Semiparametric regression modelling of nonlinear time series. *Scandinavian Journal of Statistics* 25, 521–539.

Gao, J., 2000. A semiparametric approach to pricing interest rate derivative securities. Presentation at Quantitative Methods in Finance & Bernoulli Society 2000 Conference, Sydney, 4–9 December, 2000.

Gao, J., 2004. Modelling long–range dependent Gaussian processes with application in continuous–time financial models. *Journal of Applied Probability* 41, 467–482.

Gao, J., and Anh, V., 1999. Semiparametric regression with long-range dependent error processes. *Journal of Statistical Planning & Inference* 80, 37–57.

Gao, J., and Anh, V., 2000. A central limit theorem for a random quadratic form of strictly stationary processes. *Statistics & Probability Letters* 49, 69–79.

Gao, J., Anh, V., and Heyde, C., 2002. Statistical estimation of non-stationary Gaussian processes with long–range dependence and inter-mittency. *Stochastic Processes & Their Applications* 99, 295–321.

Gao, J., Anh, V., Heyde, C., and Tieng, Q., 2001. Parameter estimation of stochastic processes with long–range dependence and intermittency. *Journal of Time Series Analysis* 22, 517–535.

Gao, J., Anh, V., and Wolff, R. C. L., 2001. Semiparametric approxi-mation methods in multivariate model selection. *Journal of Complex-ity* 17, 754–772.

Gao, J., and Gijbels, I., 2006. Selection of smoothing parametes in nonparametric and semiparametric testing. Working paper available from www.maths.uwa.edu.au/~jiti/gg51.pdf.

Gao, J., Gijbels, I., and Van Bellegem, S., 2006. Nonparametric simul-taneous testing for structural breaks. Working paper available from www.maths.uwa.edu.au/~jiti/ggvb.pdf.

Gao, J., and Hawthorne, K., 2006. Semiparametric estimation and test-ing of the trend of temperature series. *Econometrics Journal* 9, 333–356.

Gao, J., and King, M. L., 2004. Adaptive testing in continuous–time diffusion models. *Econometric Theory* 20, 844–882.

Gao, J., and King, M. L., 2005. Model estimation and specification testing in nonparametric and semiparametric models. Working paper available from www.maths.uwa.edu.au/~jiti/jems.pdf.

Gao, J., King, M. L., Lu, Z., and Tjøstheim, D., 2006. Nonparametric time series specification with nonstationarity. Working paper available from www.maths.uwa.edu.au/~jiti/gklt.pdf.

Gao, J., and Liang, H., 1995. Asymptotic normality of pseudo-LS estim-ator for partially linear autoregressive models. *Statistics & Probability Letters* 23, 27–34.

Gao, J., and Liang, H., 1997. Statistical inference in single-index and partially nonlinear regression models. *Annals of the Institute of Sta-tistical Mathematics* 49, 493–517.

Gao, J., Lu, Z., and Tjøstheim, D., 2006. Estimation in semiparametric spatial regression. *Annals of Statistics* 36, 1395–1435.

Gao, J., and Shi, P., 1997. *M*-type smoothing splines in nonparametric and semiparametric regression models. *Statistica Sinica* 7, 1155–1169.

Gao, J., and Tong, H., 2004. Semiparametric nonlinear time series model selection. *Journal of the Royal Statistical Society Series B* 66, 321–336.

Gao, J., and Tong, H., 2005. Nonparametric and semiparametric regression model selection. Unpublished technical report. Working paper available from www.maths.uwa.edu.au/~jiti/kao43.pdf.

Gao, J., Tong, H., and Wolff, R. C. L., 2002a. Adaptive orthogonal series estimation in additive stochastic regression models. *Statistica Sinica* 12, 409–428.

Gao, J., Tong, H., and Wolff, R. C. L., 2002b. Model specification tests in nonparametric stochastic regression models. *Journal of Multivariate Analysis* 83, 324–359.

Gao, J., and Wang, Q., 2006. Specification testing of nonlinear time series with long–range dependence. Working paper available from www.maths.uwa.edu.au/~jiti/gw61.pdf.

Gao, J., and Yee, T., 2000. Adaptive estimation in partially linear (semiparametric) autoregressive models. *Canadian Journal of Statistics* 28, 571–586.

Geweke, J., and Porter–Hudak, S., 1983. The estimation and application of long memory time series models. *Journal of Time Series Analysis* 4, 221–237.

Gijbels, I., and Goderniaux, A. C., 2004. Bandwidth selection for change-point estimation in nonparametric regression. *Technometrics* 46, 76–86.

González–Manteiga, W., Quintela–del–Río, A., and Vieu, P., 2002. A note on variable selection in nonparametric regression with dependent data. *Statistics & Probability Letters* 57, 259–268.

Götze, F., Tikhomirov, A., and Yurchenko, V., 2004. Asymptotic expansion in the central limit theorem for quadratic forms. Preprint 04–004, Probability and Statistics, The University of Bielefeld, Germany.

Gozalo, P. L., 1993. A consistent model specification test for nonparametric estimation of regression function models. *Econometric Theory* 9, 451–477.

Gozalo, P. L., and Linton, O. B., 2001. Testing additivity in generalized nonparametric regression models with estimated parameters. *Journal of Econometrics* 104, 1–48.

Granger, C. W. J., Inoue, T., and Morin, N., 1997. Nonlinear stochastic trends. *Journal of Econometrics* 81, 65–92.

Granger, C. W. J., and Joyeux, R., 1980. An introduction to long–range time series models and fractional differencing. *Journal of Time Series Analysis* 1, 15–30.

Granger, C. W. J., and Teräsvirta, T., 1993. *Modelling Nonlinear Dynamic Relationships.* Oxford University Press.

Granger, C. W. J., Teräsvirta, T., and Tjøstheim, D., 2006. *Nonlinear Time Series Econometrics.* Oxford University Press.

Grégoire, G., and Hamrouni, Z., 2002. Two nonparametric tests for change-point problem. *Journal of Nonparametric Statistics* 14, 87–112.

Hall, P., and Hart, J., 1990. Nonparametric regression with long–range dependence. *Stochastic Processes and Their Applications* 36, 339–351.

Hall, P., Lahiri, S., and Polzehl, J., 1996. On the bandwidth choice in nonparametric regression with both short– and long–range dependent errors. *Annals of Statistics* 23, 1921–1936.

Hannan, E. J., 1973. The asymptotic theory of linear time–series models. *Journal of Applied Probability* 10, 130–145.

Hansen, B., 2000a. Sample splitting and threshold estimation. *Econometrica* 68, 575–603.

Hansen, B., 2000b. Testing for structural change in conditional means. *Journal of Econometrics* 97, 93–115.

Härdle, W., 1990. *Applied Nonparametric Regression.* Cambridge University Press, Boston.

Härdle, W., Hall, P., and Ichimura, H., 1993. Optimal smoothing in single-index models. *Annals of Statistics* 21, 157-178.

Härdle, W., Hall, P., and Marron, J., 1988. How far are automatically chosen regression smoothing parameters from their optimum (with discussion) ? *Journal of the American Statistical Association* 83, 86–99.

Härdle, W., Hall, P., and Marron, J., 1992. Regression smoothing parameters that are not far from their optimum. *Journal of the American Statistical Association* 87, 227–233.

Härdle, W., and Kneip, A., 1999. Testing a regression model when we have smooth alternatives in mind. *Scandinavian Journal of Statistics* 26, 221–238.

Härdle, W., Liang, H., and Gao, J., 2000. *Partially Linear Models.* Springer Series: Contributions to Statistics. Physica-Verlag, New York.

Härdle, W., Lütkepohl, H., and Chen, R., 1997. A review of nonparametric time series analysis. *International Statistical Review* 65, 49–72.

Härdle, W., and Mammen, E., 1993. Comparing nonparametric versus parametric regression fits. *Annals of Statistics* 21, 1926–1947.

Härdle, W., and Vieu, P., 1992. Kernel regression smoothing of time series. *Journal of Time Series Analysis* 13, 209–232.

Hart, J., 1997. *Nonparametric Smoothing and Lack-of-Fit Tests.* Springer, New York.

Harvey, A. C., 1998. Long memory in stochastic volatility. In *Forecasting Volatility in Financial Markets* (Edited by J. Knight and S. Satchell), 307–320. Butterworth–Heinemann, Oxford.

Hengartner, N. W., and Sperlich, S., 2003. Rate optimal estimation with the integration method in the presence of many covariates. Working paper.

Heyde, C., and Gay, R., 1993. Smoothed periodogram asymptotics and estimation for processes and fields with possible long–range dependence. *Stochastic Processes and Their Applications* 45, 169–182.

Hidalgo, F. J., 1992. Adaptive semiparametric estimation in the presence of autocorrelation of unknown form. *Journal of Time Series Anal-ysis* 13, 47–78.

Hjellvik, V., and Tjøstheim, D., 1995. Nonparametric tests of linearity for time series. *Biometrika* 82, 351–368.

Hjellvik, V., Yao, Q., and Tjøstheim, D., 1998. Linearity testing using local polynomial approximation. *Journal of Statistical Planning and Inference* 68, 295–321.

Hong, Y., and Li, H., 2005. Nonparametric specification testing for continuous–time models with application to spot interest rates. *Review of Financial Studies* 18, 37–84.

Hong, Y., and White, H., 1995. Consistent specification testing via nonparametric series regression. *Econometrica* 63, 1133–1159.

Horowitz, J., 2003. Bootstrap methods for Markov processes. *Econometrica* 71, 1049–1082.

Horowitz, J. L., and Härdle, W., 1994. Testing a parametric model against a semiparametric alternative. *Econometric Theory* 10, 821–848.

Horowitz, J., and Mammen, E., 2004. Nonparametric estimation of an additive model with a link function. *Annals of Statistics* 32, 2412–2443.

Horowitz, J., and Spokoiny, V. G., 2001. An adaptive, rate-optimal test of a parametric mean-regression model against a nonparametric alternative. *Econometrica* 69, 599–632.

Huang, J., and Yang, L., 2004. Identification of non-linear additive autoregressive models. *Journal of the Royal Statistical Society Series B* 66, 463–477.

Hull, J., and White, A., 1987. The pricing of options on assets with stochastic volatilities. *Journal of Finance* 2, 281–300.

Hurst, H. E., 1951. Long–term storage capacity of reservoirs. *Transactions of the American Society of Civil Engineers* 116, 770–799.

Hurvich, C., and Beltrao, K., 1993. Asymptotics for the low-frequency ordinates of the periodogram of a long–memory time series. *Journal of Time Series Analysis* 14, 455–472.

Hurvich, C., Deo, R., and Brodsky, J., 1998. The mean squared error of Geweke and Porter-Hudak's estimator of the memory parameter of a long-memory time series. *Journal of Time Series Analysis* 19, 19–46.

Hurvich, C., and Tsai, C. L., 1995. Relative rates of convergence for efficient model selection criteria in linear regression. *Biometrika* 82, 418–425.

Hyndman, R., King, M. L., Pitrun, I., and Billah, B., 2005. Local linear forecasts using cubic smoothing splines. *Australian & New Zealand Journal of Statistics* 47, 87–99.

Jayasuriya, B., 1996. Testing for polynomial regression using nonparametric regression techniques. *Journal of the American Statistical Association* 91, 1626–1631.

Jiang, G., 1998. Nonparametric modelling of US interest rate term structure dynamics and implication on the prices of derivative securities. *Journal Financial and Quantitative Analysis* 33, 465–497.

Jiang, G., and Knight, J., 1997. A nonparametric approach to the estimation of diffusion processes with an application to a short-term interest rate model. *Econometric Theory* 13, 615–645.

Jones, C. S., 2005. Nonlinear mean reversion in the short–term interest rate. Forthcoming in the *Review of Financial Studies*.

Kantz, H., and Schreiber, T., 2004. *Nonlinear Time Series Analysis.* 2nd Edition. Cambridge University Press.

Karlsen, H., Myklebust, T., and Tjøstheim, D., 2006. Nonparametric estimation in a nonlinear cointegration model. Forthcoming in *Annals of Statistics*.

Karlsen, H., and Tjøstheim, D., 1998. Nonparametric estimation in null recurrent time series. Sonderforschungsbereich 373, 50. Humboldt University, Berlin.

Karlsen, H., and Tjøstheim, D., 2001 Nonparametric estimation in null recurrent time series. *Annals of Statistics* 29, 372–416.

Kashin, B. S., and Saakyan, A. A., 1989. *Orthogonal Series*. Translations of Mathematical Monographs, Vol. 75, American Mathematical Society.

King, M. L., and Shively, T., 1993. Locally optimal testing when a nuisance parameter is present only under the alternative. *Review of Economics and Statistics* 75 1–7.

Kohn, R., Marron, J., and Yau, P., 2000. Wavelet estimation using Bayesian basis selection and basis averaging. *Statistica Sinica* 10, 109–128.

Koul, H., and Stute, W., 1998. Regression model fitting with long memory errors. *Journal of Statistical Planning & Inference* 71, 35–56.

Kreiss, J. P., Neumann, M. H., and Yao, Q., 2002. Bootstrap tests for simple structures in nonparametric time series regression. Preprint, Institüt für Mathematische Stochastik, Technische Universität, Braunschweig, Germany.

Kristensen, D., 2004. Estimation in two classes of semiparametric diffusion models. Working paper available from Department of Economics, The University of Wisconsin–Madison, USA.

Künsch, H., 1986. Discrimination between monotonic trends and long-range dependence. *Journal of Applied Probability* 23, 1025–1030.

Lavergne, P., 2001. An equality test across nonparametric regressions. *Journal of Econometrics* 103, 307–344.

Lavergne, P., and Vuong, Q. H., 1996. Nonparametric selection of regressors: the nonnested case. *Econometrica* 64, 207–219.

Lavergne, P., and Vuong, Q. H., 2000. Nonparametric significance testing. *Econometric Theory* 16, 576–601.

Li, K. C., 1985. From Stein's unbiased risk estimates to the method of generalized cross-validation. *Annals of Statistics* 13, 1352–1377.

Li, K. C., 1986. Asymptotic optimality of C_L and generalized cross-validation in ridge regression with application to spline smoothing. *Annals of Statistics* 14, 1101–1112.

Li, K. C., 1987. Asymptotic optimality for C_p, C_L, cross-validation and generalized cross-validation: discrete index set. *Annals of Statistics* 15, 958–975.

Li, M., Pearson, N. D., and Poteshman, A. M., 2005. Facing up to conditioned diffusions. Forthcoming in the *Journal of Financial Economics*.

Li, Q., 1999. Consistent model specification tests for time series econometric models. *Journal of Econometrics* 92, 101–147.

Li, Q., and Hsiao, C., 1998. Testing serial correlation in semiparametric panel data models. *Journal of Econometrics* 87, 207–237.

Li, Q., Hsiao, C., and Zinn, J., 2003. Consistent specification tests for semiparametric & nonparametric models based on series estimation methods. *Journal of Econometrics* 112, 295–325.

Li, Q., and Racine, J., 2006. *Nonparametric Econometrics: Theory and Practice.* Princeton University Press.

Li, Q., and Wang, S., 1998. A simple consistent bootstrap tests for a parametric regression functional form. *Journal of Econometrics* 87, 145–165.

Li, Q., and Wooldridge, J., 2002. Semiparametric estimation of partially linear models for dependent data with generated regressors. *Econometric Theory* 18, 625–645.

Lieberman, O., and Phillips, P. C. B., 2004. Expansions for the distribution of the maximum likelihood estimator of the fractional difference parameter. *Econometric Theory* 20, 464–484.

Lieberman, O., and Phillips, P. C. B., 2005. Expansions for approximate maximum likelihood estimators of the fractional difference parameter. *Econometrics Journal* 8, 367–379.

Ling, S., and Tong, H., 2005. Testing for a linear MA model against threshold MA models. *Annals of Statistics* 33, 2529–2552.

Linton, O. B., 1997. Efficient estimation of additive nonparametric regression models. *Biometrika* 84, 469–473.

Linton, O. B., 2000. Efficient estimation of generalized additive nonparametric regression models. *Econometric Theory* 16, 502–523.

Linton, O. B., 2001. Estimating additive nonparametric models by partial L_q norm: the curse of fractionality. *Econometric Theory* 17, 1037–1050.

Linton, O. B., and Härdle, W., 1996. Estimation of additive regression models with known links. *Biometrika* 83, 529–540.

Linton, O. B., and Mammen, E., 2005. Estimation in semiparametric ARCH(∞) models by kernel smoothing methods. *Econometrica* 73, 1001–1030.

Mammen, E., Linton, O. B., and Nielsen, J. P., 1999. The existence and asymptotic properties of a backfitting projection algorithm under weak conditions. *Annals of Statistics* 27, 1443–1490.

Mandelbrot, B., and Van Ness, J., 1968. Fractional Brownian motion, fractional noises and applications. *SIAM Review* 10, 422–437.

Masry, E., and Tjøstheim, D., 1995. Nonparametric estimation and identification of nonlinear ARCH time series. *Econometric Theory* 11, 258–289.

Masry, E., and Tjøstheim, D., 1997. Additive nonlinear ARX time series and projection estimates. *Econometric Theory* 13, 214–252.

Merton, R. C., 1973. The theory of rational option pricing. *Bell Journal of Economics*, 4, 141–183.

Mikosch, T., and Starica, C., 2004. Nonstationarities in financial time series, the long–range dependence and the IGARCH effects. *Review of Economics and Statistics*, 86, 378–390.

Nicolau, J., 2003. Bias reduction in nonparametric diffusion coefficient estimation. *Econometric Theory* 19, 754–777.

Nielsen, J. P., and Linton, O. B., 1998. An optimization interpretation of integration and back–fitting estimators for separable nonparametric models. *Journal of the Royal Statistical Society Series B* 60, 217–222.

Nishiyama, Y., and Robinson, P., 2000. Edgeworth expansions for semiparametric averaged derivatives. *Econometrica* 68, 931–980.

Nishiyama, Y., and Robinson, P., 2005. The bootstrap and the Edgeworth correction for semiparametric averaged derivatives. *Econometrica* 73, 903–948.

Pagan, A., and Ullah, A., 1999. *Nonparametric Econometrics*. Cambridge University Press, New York.

Park, J., and Phillips, P. C. B., 2001. Nonlinear regressions with integrated time series. *Econometrica* 69, 117–162.

Phillips, P. C. B., 1987. Time series regression with a unit root. *Econometrica* 55, 277–302.

Phillips, P. C. B., and Park, J., 1998. Nonstationary density estimation and kernel autoregression. Cowles Foundation Discussion Paper, No. 1181, Yale University.

Pollard, D., 1984. *Convergence of Stochastic Processes*. Springer, New York.

Pritsker, M., 1998. Nonparametric density estimation and tests of continuous time interest rate models. *Review of Financial Studies* 11, 449–487.

Robinson, P. M., 1988. Root–N–consistent semiparametric regression. *Econometrica* 56, 931–964.

Robinson, P. M., 1989. Hypothesis testing in semiparametric and nonparametric models for econometric time series. *Review of Economic Studies* 56, 511–534.

Robinson, P. M., 1994. Time series with strong dependence. In *Advances in Econometrics* (Edited by C. A. Sims), Sixth World Congress Vol. 1, 47–96. Cambridge University Press, Cambridge.

Robinson, P. M., 1995a. Log-periodogram regression of time series with long-range dependence. *Annals of Statistics* 23, 1048–1072.

Robinson, P. M., 1995b. Gaussiaan semiparametric estimation of long–range dependence. *Annals of Statistics* 23, 1630–1661.

Robinson, P. M., 1997. Large-sample inference for nonparametric regression with dependent errors. *Annals of Statistics* 25, 2054–2083.

Robinson, P. M., 2001. The memory of stochastic volatility models. *Journal of Econometrics* 101, 195–218.

Robinson, P. M. (ed.), 2003. Time series with long memory. *Advanced Texts in Econometrics*. Oxford University Press, Oxford.

Robinson, P. M., 2005. The distance between rival nonstationarity fractional processes. *Journal of Econometrics* 128, 283–300.

Robinson, P. M., and Zaffaroni, P., 1998. Nonlinear time series with long memory: a model for stochastic volatility. *Journal of Statistical Planning & Inference* 68, 359–371.

Rockafeller, R. T., 1970. *Convex Analysis*. Princeton University Press, New Jersey.

Roussas, G., and Ioannides, D., 1987. Moment inequalities for mixing sequences of random variables. *Stochastic Analysis and Applications* 5, 61–120.

Ruppert, D., Wand, M. P., and Carroll, R. J., 2003. *Semiparametric Regression*. Cambridge University Press, Cambridge.

Samarov, A. M., 1993. Exploring regression structure using nonparametric functional estimation. *Journal of the American Statistical Association* 423, 836–847.

Schumaker, L., 1981. *Spline Functions*. John Wiley, New York.

Shao, J., 1993. Linear model selection by cross–validation. *Journal of the American Statistical Association* 422, 486–494.

Shao, J., 1997. An asymptotic theory for linear model selection (with comments). *Statistica Sinica* 7, 221–264.

Shi, P., and Tsai, C. L., 1999. Semiparametric regression model selections. *Journal of Statistical Planning & Inference* 77, 119–139.

Shimotsu, K., and Phillips, P. C. B., 2005. Exact local Whittle estimation of fractional integration. *Annals of Statistics* 33, 1890–1933.

Shimotsu, K., and Phillips, P. C. B., 2006. Local Whittle estimation of fractional integration and some of its variants. *Journal of Econometrics* 130, 209–233.

Shiryaev, A. N., 1999. *Essentials of Stochastic Finance*. World Scientific, Singapore.

Shively, T., and Kohn, R., 1997. A Bayesian approach to model selection in stochastic coefficient regression models and structural time series models. *Journal of Econometrics* 76, 39–52.

Shively, T., Kohn, R., and Ansley, C. F., 1994. Testing for linearity in a semiparametric regression model. *Journal of Econometrics* 64, 7–96.

Shively, T., Kohn, R., and Wood, S., 1999. Variable selection and function estimation in additive nonparametric regression using a data-based prior. With comments and a rejoinder by the authors. *Journal of the American Statistical Association* 94, 777–806.

Silverman, B. W., 1986. *Density Estimation for Statistics and Data Analysis*. Chapman and Hall, London.

Sperlich, S., Tjøstheim, D., and Yang, L., 2002. Nonparametric estimation and testing of interaction in additive models. *Econometric Theory* 18, 197–251.

Stanton, R., 1997. A nonparametric model of term structure dynamics and the market price of interest rate risk. *Journal of Finance* 52, 1973–2002.

Stock, J., 1994. Unit roots, structural breaks and trends. In *Handbook of Econometrics* (Edited by R. Engle and D. McFadden) 4, 2740–2841. Elsevier Science, Amsterdam.

Stone, M., 1977. An asymptotic equivalence of choice of model by cross-validation and Akaike's criterion. *Journal of the Royal Statistical Society Series B* 39, 44–47.

Stute, W., 1997. Nonparametric model checks for regression. *Annals of Statistics* 25, 613–641.

Stute, W., Thies, S., and Zhu, L., 1998. Model checks for regression: an innovation process approach. *Annals of Statistics* 26, 1916–1934.

Stute, W., and Zhu, L., 2002. Model checks for generalized linear models. *Scandinavian Journal of Statistics* 29, 535–545.

Stute, W., and Zhu, L., 2005. Nonparametric checks for single-index models. *Annals of Statistics* 33, 1048–1083.

Sun, Y., and Phillips, P. C. B., 2003. Nonlinear log–periodogram regression for perturbed fractional processes. *Journal of Econometrics* 115, 355–389.

Sundaresan, S., 2001. Continuous–time methods in finance: a review and an assessment. *Journal of Finance* 55, 1569–1622.

Tanaka, K., 1996. *Time Series Analysis: Nonstationary and Noninvertible Distribution Theory*. John Wiley & Sons, New York.

Taylor, S., 1986. *Modelling Financial Time Series*. John Wiley, Chichester. U.K.

Taylor, S. J., 1994. Modelling stochastic volatility: a review and comparative study. *Mathematical Finance* 4, 183–204.

Teräsvirta, T., Tjøstheim, D., and Granger, C., 1994. Aspects of modelling nonlinear time series. In *Handbook of Econometrics* (Edited by R. F. Engle and D. L. McFadden) 4, 2919–2957. Elsevier Science, Amsterdam.

Tjøstheim, D., 1994. Nonlinear time series: a selective review. *Scandinavian Journal of Statistics* 21, 97–130.

Tjøstheim, D., 1999. Nonparametric specification procedures for time series. *Asymptotics, nonparametrics, and time series* 158, 149–199. Statistics: Textbooks and Monographs. Dekker, New York.

Tjøstheim, D., and Auestad, B., 1994a. Nonparametric identification of nonlinear time series: projections. *Journal of the American Statistical Association* 89, 1398–1409.

Tjøstheim, D., and Auestad, B., 1994b. Nonparametric identification of nonlinear time series: selecting significant lags. *Journal of the American Statistical Association* 89, 1410–1419.

Tong, H., 1976. Fitting a smooth moving average to noisy data. *IEEE Transactions on Information Theory*, IT-26, 493–496.

Tong, H., 1990. *Nonlinear Time Series*. Oxford University Press, Oxford.

Truong, Y. K., and Stone, C. J., 1994. Semiparametric time series regression. *Journal Time Series Analysis* 15, 405–428.

Tsay, R., 2005. *Analysis of Financial Time Series*. Second Edition. Wiley Interscience, Hoboken, New Jersey.

Tschernig, R., and Yang, L., 2000. Nonparametric lag selection for time series. *Journal of Time Series Analysis* 21, 457–487.

Vasicek, O., 1977. An equilibrium characterization of the term structure. *Journal of Financial Economics*, 5, 177–188.

Viano, M., Deniau, C., and Oppenheim, G., 1994. Continuous–time fractional ARMA processes. *Statistics & Probability Letters* 21, 323–336.

Viano, M., Deniau, C., and Oppenheim, G., 1995. Long-range dependence and mixing for discrete time fractional processes. *Journal of Time Series Analysis* 16, 323–338.

Vieu, P., 1994. Choice of regressors in nonparametric estimation. *Computational Statistics & Data Analysis* 17, 575–594.

Vieu, P., 1995. Order choice in nonlinear autoregressive models. *Statistics* 26, 307–328.

Vieu, P., 2002. Data–driven model choice in multivariate nonparametric regression. *Statistics* 36, 231–245.

Wahba, G., 1978. Improper priors, spline smoothing and the problem of guarding against model errors in regression. *Journal of the Royal Statistical Society Series B* 40, 364–372.

Wahba, G., 1990. *Spline Models for Observational Data*. SIAM, Philadelphia.

Wand, M. P., and Jones, M. C., 1995. *Kernel Smoothing*. Chapman and Hall, London.

Whang, Y. J., 2000. Consistent bootstrap tests of parametric regression functions. *Journal of Econometrics* 98, 27–46.

Whang, Y. J., and Andrews, D. W. K., 1993. Tests of specification for parametric and semiparametric models. *Journal of Econometrics* 57, 277–318.

Wong, C. M., and Kohn, R., 1996. A Bayesian approach to esti-mating and forecasting additive nonparametric autoregressive models. *Journal of Time Series Analysis* 17, 203–220.

Wong, F., Carter, C., and Kohn, R., 2003. Efficient estimation of co-variance selection models. *Biometrika* 90, 809–830.

Wood, S., Kohn, R., Shively, T., and Jiang, W., 2002. Model selection in spline nonparametric regression. *Journal of the Royal Statistical Society Series B* 64, 119–139.

Wooldridge, J., 1992. A test for functional form against nonparametric alternatives. *Econometric Theory* 8, 452–475.

Xia, Y., Li, W. K., Tong, H., and Zhang, D., 2004. A goodness-of-fit test for single-index models (with comments). *Statistica Sinica* 14, 1–39.

Xia, Y., Tong, H., and Li, W. K., 1999. On extended partially linear single–index models. *Biometrika* 86, 831–842.

Xia, Y., Tong, H., Li, W. K., and Zhu, L. X., 2002. An adaptive esti-mation of dimension reduction space. *Journal of the Royal Statistical Society Series B* 64, 363–410.

Yang, L., 2002. Direct estimation in an additive model when the components are proportional. *Statistica Sinica* 12, 801–821.

Yang, L., 2006. A semiparametric GARCH model for foreign exchange volatility. *Journal of Econometrics* 130, 365–384.

Yang, L., and Tschernig, R., 2002. Non– and semiparametric identification of seasonal nonlinear autoregression models. *Econometric Theory* 18, 1408–1448.

Yang, Y., 1999. Model selection for nonparametric regression. *Statistica Sinica* 9, 475–500.

Yao, Q., and Tong, H., 1994. On subset selection in nonparametric stochastic regression. *Statistica Sinica* 4, 51–70.

Yatchew, A. J., 1992. Nonparametric regression tests based on least squares. *Econometric Theory* 8, 435–451.

Yau, P., and Kohn, R., 2003. Estimation and variable selection in nonparametric heteroscedastic regression. *Statistics & Computing* 13, 191–208.

Yau, P., Kohn, R., and Wood, S., 2003. Bayesian variable selection and model averaging in high-dimensional multinomial nonparametric regression. *Journal of Computational & Graphical Statistics* 12, 23–54.

Zhang, C. M., and Dette, H., 2004. A power comparison between nonparametric regression tests. *Statistics & Probability Letters* 66, 289–301.

Zhang, P., 1991. Variable selection in nonparametric regression with continuous covariates. *Annals of Statistics* 19, 1869–1882.

Zhang, P., 1993. Model selection via multifold cross–validation. *Annals of Statistics* 21, 299–313.

Zheng, J. X., 1996. A consistent test of functional form via nonparametric estimation techniques. *Journal of Econometrics* 75, 263–289.

Zheng, X., and Loh, W. Y., 1995. Consistent variable selection in linear models. *Journal of the American Statistical Association* 90, 151–156.

Zheng, X., and Loh, W. Y., 1997. A consistent variable selection criterion for linear models with high–dimensional covariates. *Statistica Sinica* 7, 311–325.

Zhu, L., 2005. *Nonparametric Monte Carlo Tests and Their Applications*. Lecture Notes in Statistics. 182. Springer, New York.

Author Index

Karlsen, H., 9, 48
Karolyi, F., 112, 113, 131, 137, 144, 145
Kasahara, Y., 171
Kashin, B. S., 41
King, M. L., 6, 8, 9, 17, 48, 49, 56, 66, 69, 71, 80, 81, 154, 156, 157, 169, 191, 193
Knight, J., 113, 114, 120, 121, 145, 156
Kohn, R., 25, 35, 80, 110
Kokoszka, P., 192
Koul, H., 192
Kreiss, J. P., 53, 59, 80
Kristensen, D., 10, 115, 124, 125, 156

Lütkepohl, H., 6, 83
Lahiri, S., 192
Lavergne, P., 69
Li, H., 50, 63, 66, 115, 156
Li, K. C., 19, 21, 41
Li, M., 48, 54, 55, 80, 156
Li, Q., 6, 9, 14, 47, 48, 55, 56, 58–60, 69, 70, 80, 127, 129, 151, 153, 193
Li, R., 17, 48, 56, 69, 110
Li, W. K., 8, 17, 36, 38, 43, 47, 48, 70
Liang, H., 2, 3, 6, 8, 14, 16, 17, 48, 69, 70, 80, 83, 89, 110
Lieberman, O., 192
Ling, S., 81
Linton, O., 17, 47, 48, 51, 55, 56, 58, 59, 71, 80
Linton, O. B., 47
Liu, J., 80
Lo, A., 156
Loh, W. Y., 92, 94, 103, 105
Longstaff, F., 112, 113, 131, 137, 144, 145

Lorentz, G. G., 41
Lu, Z., 9, 16, 43, 45, 47, 81
Lund, J., 169

Mammen, E., 17, 42, 48, 49, 52, 59, 80, 127, 129
Mandelbrot, B., 157, 192
Marron, J., 21, 110
Masry, E., 22, 95
Merton, R. C., 111, 112
Mikosch, T., 4
Morin, N., 9, 48
Moscatelli, V. B., 48
Myklebust, T., 9, 48

Neumann, M. H., 53, 59, 80
Nicolau, J., 114
Nielsen, J. P., 17
Nishiyama, Y., 51

Oppenheim, G., 192

Pagan, A., 14
Park, J., 9, 48
Pearson, N., 156
Phillips, P. C. B., 9, 10, 48, 114, 156, 192
Pitrun, I., 48
Ploberger, W., 81
Pollard, D., 199
Polzehl, J., 192
Porter–Hudak, S., 157, 159, 192
Poteshman, A. M., 156
Pritsker, M., 49, 132, 156

Quintela–del–Río, A., 69

Racine, J., 6, 14
Renault, E., 11, 164, 168, 170
Rice, J. A., 2
Robinson, P. M., 6, 51, 80, 157, 159–161, 169, 176, 191, 192
Rockafeller, R. T., 167

Wolff, R. C. L., 6, 16, 19, 22, 25, 43, 48, 71, 80, 84

Wong, C. M., 25, 35, 110

Wood, S., 110

Wooldridge, J., 47, 80

Xia, Y., 8, 17, 36, 38, 43, 47, 48, 70

Yang, L., 15, 48, 71, 81, 110

Yang, Y., 110

Yao, Q., 9, 14, 35, 41, 48–50, 53, 59, 80, 81, 83, 85, 92, 94, 110, 127, 193

Yatchew, A. J., 80

Yau, P., 110

Yee, T., 6, 7

Yurchenko, V., 75, 196, 197, 204

Zaffaroni, P., 192

Zhang, C. M., 50, 51, 56, 66, 80, 81, 114, 156

Zhang, D., 8, 70

Zhang, J., 51, 56

Zhang, P., 85, 106, 110

Zheng, J. X., 55, 56, 58, 80

Zheng, X., 92, 94, 103, 105

Zhou, Z., 156

Zhu, L., 48, 50, 70

Zinn, J., 80

Subject Index

T - #0414 - 071024 - C3 - 229/152/11 - PB - 9780367389352 - Gloss Lamination